Lecture Notes in Networks and Systems

Volume 79

The series "Lecture Notes in Networks and Systems" publishes the latest developments in Networks and Systems—quickly, informally and with high quality. Original research reported in proceedings and post-proceedings represents the core of LNNS.

Volumes published in LNNS embrace all aspects and subfields of, as well as new challenges in, Networks and Systems.

The series contains proceedings and edited volumes in systems and networks, spanning the areas of Cyber-Physical Systems, Autonomous Systems, Sensor Networks, Control Systems, Energy Systems, Automotive Systems, Biological Systems, Vehicular Networking and Connected Vehicles, Aerospace Systems, Automation, Manufacturing, Smart Grids, Nonlinear Systems, Power Systems, Robotics, Social Systems, Economic Systems and other. Of particular value to both the contributors and the readership are the short publication timeframe and the world-wide distribution and exposure which enable both a wide and rapid dissemination of research output.

The series covers the theory, applications, and perspectives on the state of the art and future developments relevant to systems and networks, decision making, control, complex processes and related areas, as embedded in the fields of interdisciplinary and applied sciences, engineering, computer science, physics, economics, social, and life sciences, as well as the paradigms and methodologies behind them.

**** Indexing: The books of this series are submitted to ISI Proceedings, SCOPUS, Google Scholar and Springerlink ****

More information about this series at http://www.springer.com/series/15179

Jyotsna Kumar Mandal · Kallol Bhattacharya ·
Ivy Majumdar · Surajit Mandal
Editors

Information, Photonics and Communication

Proceedings of Second National Conference, IPC 2019

Springer

Editors
Jyotsna Kumar Mandal
Department of Computer Science
and Engineering
University of Kalyani
Kalyani, West Bengal, India

Kallol Bhattacharya
University of Calcutta
Kolkata, West Bengal, India

Ivy Majumdar
B.P. Poddar Institute of Management
and Technology
Kolkata, West Bengal, India

Surajit Mandal
B.P. Poddar Institute of Management
and Technology
Kolkata, West Bengal, India

ISSN 2367-3370 ISSN 2367-3389 (electronic)
Lecture Notes in Networks and Systems
ISBN 978-981-32-9452-3 ISBN 978-981-32-9453-0 (eBook)
https://doi.org/10.1007/978-981-32-9453-0

This Springer imprint is published by the registered company Springer Nature Singapore Pte Ltd.
The registered company address is: 152 Beach Road, #21-01/04 Gateway East, Singapore 189721, Singapore

Preface

This volume entitled Information, Photonics and Communication (IPC'19) contains the Proceedings of Second National Conference organized by the Department of Electronics and Communication Engineering, B.P. Poddar Institute of Management and Technology at Kolkata, India, during 1–3 February 2019. The proceedings are published in Lecture Notes in Networks and Systems, Springer.

The book covers multiple domains under categories of microelectronics and VLSI, communication systems and network and signal processing and computational intelligence. Wireless and mobile communication, signal, image and video processing, DSP algorithm and architecture, optical networks, devices and materials, embedded and VLSI system design are the main contents of the book.

There are four articles in microelectronics and VLSI section. These are computation of gate-induced drain leakage current, investigating the effect of structural parameters on static characteristics of ultrathin DG MOSFET, implementation of L-shaped dielectric double metal dual-gate TFET and memristor-based circuit design.

In communication system and network section, the articles are on performance analysis of AntHocNet, AND-OR-Invert logic for photonic integrated circuits, identifying influential nodes based on network topology, cross-media encryption-cum-obfuscation technique, computing electromagnetic bandgap structure, inter-vehicular information system (IVIS) on IoT platform and generation of renewable electrical energy from noise.

In signal processing and computational intelligence section, the articles are on mining user's data based on customer's rating for prediction and recommendation, Charlier and Meixner moments and their application for texture and image de-noising problems, tumour boundary delineation, detection and classification of cervical spondylosis, visual saliency-based video summarization, moment-based feature extraction for texture image retrieval, compression of satellite images using Butterworth filtering and advanced wavelet transform for image processing.

The editors express their gratitude to Springer Nature authority for publishing this volume in Springer.

The editors also express sincere thanks to the reviewers for their dedications in reviewing the articles. Also, thanks to the authors for submitting their articles into this volume.

Hope this volume will be a good reference manual for researchers and budding engineers.

West Bengal, India

Jyotsna Kumar Mandal
Kallol Bhattacharya
Ivy Majumdar
Surajit Mandal
Editors

Contents

Microelectronics and VLSI

Computation of Gate-Induced-Drain-Leakage Current Due to Band-to-Band Tunneling for Ultrathin MOSFET 3
Krishnendu Roy, Anal Roy Chowdhury and Arpan Deyasi

Investigating Effect of Structural Parameters on Static Characteristics of Ultrathin DG MOSFET Using Taur's Model 11
Riya Chakraborty, Deepanwita Mondal and Arpan Deyasi

Implementation of L-Shaped Dielectric Double Metal Dual-Gate TFET Toward Improved Performance Characteristics and Reduced Ambipolarity 21
Bijoy Goswami, Sutanni Bhowmick, Arindam Haldar, Goutika Paul, Debadipta Basak and Subir Kumar Sarkar

Memristor-Based Circuit Design Approaches and Future Challenges 31
Ramesh Kumar and Surajit Mandal

Communication System and Network

Performance Analysis of AntHocNet Based on NS2 45
Khondekar Lutful Hassan and Jyotsna Kumar Mandal

Design and Simulation of AND-OR-INVERT Logic for Photonic Integrated Circuits 55
Mahesh V. Sonth, Sanjaykumar Gowre, Nagshettappa Biradar and Basavaraj Gadgay

Identifying Influential Nodes Based on Network Topology: A Comparative Study 65
Anindita Raychaudhuri, Subhasis Mallick, Ankit Sircar and Shalini Singh

Towards a Novel Cross-media Encryption-Cum-Obfuscation Technique 77
Dipnarayan Das and Sumit Gupta

Computing Electromagnetic Bandgap Structure of Metamaterial-Based 2D Photonic Crystal for TM Mode 87
Ratul Ghosh, Papri Chakraborty, Anwesha Adhikary and Arpan Deyasi

Inter-Vehicular Information System (IVIS) on IOT Platform 95
Soumyadip Chatterjee

A Technique for Generation of Renewable Electrical Energy from Noise 101
Soumyadip Sarkar, Aheli Das and Arijit Saha

Signal Processing and Computational Intelligence

Mining User's Data Based on Customer's Rating for Prediction and Recommendation—A Comparative Analysis 109
Soma Bandyopadhyay, S. S. Thakur and Jyotsna Kumar Mandal

Charlier and Meixner Moments and Their Application for Texture and Image De-noising Problems 123
Perugu Ananth Raj

Tumor Boundary Delineation Using Abnormality Outlining Box Guided Modified GVF Snake Model 135
Srinivas Thirumala and Srinivasa Rao Chanamallu

Detection and Classification of Cervical Spondylosis Using Image Segmentation Techniques 145
Aniruddha Paul, Aritra Paul and Pramit Brata Chanda

Visual Saliency Based Video Summarization: A Case Study For Preview Video Generation 155
G. Ramya and Subhash Kulkarni

A Moment Based Feature Extraction for Texture Image Retrieval 167
Ivy Majumdar, B. N. Chatterji and Avijit Kar

A Novel Approach to Compression of Satellite Images Using Butterworth Filtering 179
Anirban Patra, Swagata Bandyopadhyay, Debasish Chakraborty and Arijit Saha

Advanced Wavelet Transform for Image Processing—A Survey 185
Manas Saha, Mrinal Kanti Naskar and B. N. Chatterji

Classification of RBC and WBC in Noisy Microscopic Images of Blood Smear 195
Sayantari Ghosh and Saumik Bhattacharya

Impact of Curvature on Intensity-Based Non-rigid Medical Image Registration 201
Prasenjit Kumar Mudi

Signal Processing and Nuclear Structure of ^{106}Cd Nucleus 219
D. Choudhury and R. Goswami

Performance Study of Some Recent Optimization Techniques for Energy Minimization in Surveillance Video Synopsis Framework 227
Subhankar Ghatak and Suvendu Rup

Author Index 239

About the Editors

Jyotsna Kumar Mandal M.Sc. (Ph.), JU, M.Tech. (CS), CU, Ph.D. (Eng.), JU, Professor CSE, former dean of FETM, KU for two consecutive terms. He has teaching and research experience spanning 30 years and has completed four AICTE projects and one state govt. project. He is a life member of the CSI, CRSI, a member of the ACM, and fellow of IETE. He was also honorary vice chairman and chairman of the CSI. He has delivered over 100 lectures and organized more than 30 national and international conferences. He is an editorial board member and corresponding editor of the Proceedings of Science Direct, IEEE and other conferences as well as guest editor of the MST Journal. He has published more than 400 research articles and six books. Twenty-three scholars were awarded under his supervision.

Kallol Bhattacharya is a Professor at the Department of Applied Optics & Photonics, University of Calcutta. He completed his B.Sc., M.Sc. and Ph.D. from the Department of Applied Physics, University of Calcutta in 1985, 1988, 1996 respectively, and has worked as a lecturer at Jadavpur University and Scientist D at Saha Institute of Nuclear Physics. His research interests include optical metrology, polarization interferometry, birefringence measurements, nonlinear optics and optical coherence tomography, and he has published numerous papers in these areas. He has received a Monbusho fellowship from the Japanese government and a research award from Mitutoyo Association of Science and Technology, Japan.

Ivy Majumdar is an Associate Professor at the Department of Electronics and Communication Engineering, B.P. Poddar Institute of Management and Technology. She has completed her B.Sc., B.Tech. and M.Tech. in Physics from the University of Calcutta and Ph.D. from Jadavpur University in Engineering. Her research interests include signal and image processing. She has published around 10 journal and conference papers and 2 book chapters. She has also implemented a solar panel in the slum areas of Sundarban, Domjur, and Sodepur.

Surajit Mandal is an Associate Professor at B.P. Poddar Institute of Management & Technology, Kolkata, India. He completed his B.Sc. in Physics from Krishna College, Murshidabad, and his M.Tech. in Optics & Optoelectronics and Ph.D. in Applied Optics and Photonics from the University of Calcutta. His research interests include optical and electronic signal and image processing, image evaluation, signal analysis and conditioning, cryptography and steganography, and he has published numerous papers in these areas in national and international conferences and journals. He is also a reviewer of several international journals, like Optics Express, Optics Letters, Applied Optics, JOSA, Optik, African Journal of Physics and Journal of Engineering and Technology Research. He has also received the 2017 'Teacher of the year' award from Maulana Abul Kalam Azad University of Technology, West Bengal.

Microelectronics and VLSI

Computation of Gate-Induced-Drain-Leakage Current Due to Band-to-Band Tunneling for Ultrathin MOSFET

Krishnendu Roy, Anal Roy Chowdhury and Arpan Deyasi

Abstract In this paper, gate-induced-drain-leakage current due to band-to-band tunneling is analytically computed in nanometric MOSFET under high electric field. Fowler-Nordheim tunneling current is first calculated for different dielectric thicknesses with some alteration of Hu's model, and dominance of thermionic current is established. Under this criterion, B–B tunneling current is evaluated as leakage arises due to the overlap of gate over source and drain regions, and dielectric properties along with doping concentration and temperature are taken into account following Kane's tunneling probability. Role of high-*K* dielectric material is also analyzed for nanoscale application. Result shows that higher overlapping of gate length due to lateral diffusion of source and drain regions tailors the leakage current. Findings are extremely important for use of the device as SRAM.

Keywords Gate-induced-drain-leakage · Gate overlap length · High-*K* dielectric · Band-to-band tunneling · Doping concentration · Dielectric thickness

1 Introduction

Nanoelectronics Scaling of MOSFET following Moore's law started almost three decades ago and becomes subject of interest due to the novel features offered by low-dimensional structures, and the technology roadmap [1] published in 2007 speaks in favor of further shrinkage keeping in mind of ever-increasing problems of short-channel effect [2, 3] . Effective gate control, henceforth, becomes a prime

K. Roy · A. R. Chowdhury
Department of Electronic Science, A.P.C College, Barasat, India
e-mail: krishnendu.physics94@gmail.com

A. R. Chowdhury
e-mail: analroychowdhury084@gmail.com

A. Deyasi (✉)
Department of Electronics and Communication Engineering, RCC Institute of Information Technology, Kolkata, India
e-mail: deyasi_arpan@yahoo.co.in

J. K. Mandal et al. (eds.), *Information, Photonics and Communication*,
Lecture Notes in Networks and Systems 79,
https://doi.org/10.1007/978-981-32-9453-0_1

research topic in the days now [4, 5], and computation of tunneling current gets also importance in this regard. Several works are already reported on direct tunneling occurs due to both vertical and lateral electric field [6–8] following the fundamental physics provided by Fowler-Nordheim [9] . Structural parameters of the device [10] and material compositions in the dielectric layer [11, 12] play crucial role in this context for estimation of tunneling current. Inverted Si surface makes way in favor of quantum-mechanical tunneling as proved by Lai [13], following the pioneering work of Chang [14] on double-gate structures. Hu [15] added a correction factor on Chang's work to compute the effect for ultrathin dielectrics. But these reports exclude the effect of band-to-band tunneling.

Under depletion condition, due to overlapping of gate over the source and drain regions [occurs due to lateral diffusion], band bending occurs, and it is maximum at the interface. It can be tuned by varying drain-to-gate potential difference. If the bending becomes very large, then valence electrons near the edge of the band tunnels to the empty energy levels of the conduction band, and the process provides a significant output when doping concentration of the substrate is high. These electrons are collected at drain end, and holes are directed toward the channel, thus maintaining depletion condition [16, 17]. The tunneling current arises and exists for both positive and negative vertical fields. Chen assumed that doping concentration remains almost constant throughout the overlapped region [18], and considering that its effect on the overall leakage of the VLSI circuit is estimated [19], which is useful for memory applications. Choi computed it for ultrathin SGMOSFET and DGMOSFET [20] and then hot carrier effect is added [21].

In the present paper, effect of dielectric thickness and relative permittivity of the insulating layer is analytically estimated on leakage current with varying doping concentration and gate overlapped region. Temperature effect is also investigated for both moderate and high bias ranges. Computation is carried out for moderate and high fields, as overlap of gate causes higher electric field, and simultaneously larger band bending. Results are extremely useful for memory applications.

2 Mathematical Formulation

The direct-tunneling current density for a dielectric voltage V_{ox} is smaller than the barrier height Φ_b. Corresponding tunneling current is given by Eq. (1) as given by Hu [15]:

$$J_n = \frac{q^3}{8\pi h\phi_b\varepsilon_{\text{ox}}} C(V_G, V_{\text{ox}}, T_{\text{ox}}, \phi_b) \exp\left[-8\pi\sqrt{2m^*_{\text{ox}}\phi_b^{1.5}}\left(1-\left(1-\frac{V_{\text{ox}}}{\phi_b}\right)^{1.5}\right)/3hqE_{\text{ox}}\right] \quad (1)$$

where

$$C = \exp\left[\left(\frac{20}{\phi_b}\right)\left(\frac{(|V_{\text{ox}}|-\phi_b)}{\phi_{b0}}+1\right)^{\alpha}\left(1-\frac{V_{\text{ox}}}{\phi_b}\right)\right]\left(\frac{V_G}{T_{\text{ox}}}\right)N \quad (2)$$

where 'C' becomes a constant for a given range of electric field, and thus, the variation of 'J' is very small. At low-dimensional device structure, current density varies significantly. The present authors tailor the expression of constant 'C' which when fits in the Moore's law gives

$$C = \exp\left[\left(\frac{K}{\phi_b}\right)\left(\frac{(|V_{\text{ox}}| - \phi_b)}{\phi_{b0}} + 1\right)^{\alpha}\left(1 - \frac{V_{\text{ox}}}{\phi_b}\right)\right]\left(\frac{V_G}{T_{\text{ox}}}\right)N \tag{3}$$

where 'K' varies with small change of electric field.

Band-to-band current density is given by

$$I_{\text{GIDL}} = W\ell_{\text{ov}} \int_0^{\infty} qT(\xi_{\text{si}})\mathrm{d}x \tag{4}$$

where 'T' gives the Kane's tunneling probability. Substituting the values of the probability and assuming the constancy of doping concentration over the entire range of gate-shadowed doped regions, Eq. (4) is modified in the following form

$$I_{\text{GIDL}} = \frac{AW\ell_{\text{ov}}\varepsilon E^4(0)}{\xi_{\text{si}}ND}\exp\left(-\frac{\xi_{\text{si}}}{E(0)}\right) \tag{5}$$

where 'A' and ξ_{si} are the fitting parameters dependent on bandgap and effective mass of the substrate. The bandgap is also temperature dependent.

3 Results and Discussion

Based on the Eq. (5), Fowler-Nordheim tunneling current is computed in presence of high electric field and result is graphically represented. The lateral high field causes leakage current in the device. The magnitude may be negligible under low bias condition, but increases with increase of field intensity. This is depicted in Fig. 1 for two different values of V_{DS}. Result is calculated for different dielectric thickness again, and lower leakage current is observed for higher thickness. It is observed that for negative gate voltage, leakage current is substantially higher due to increase of surface field. But the magnitude of current is negligibly small for low-to-moderate values of V_{DS}, seen in Fig. 1a and b, respectively. For large horizontal field, leakage current becomes non-negligible due to rapid growth of thermionic current and thus affects the device performance.

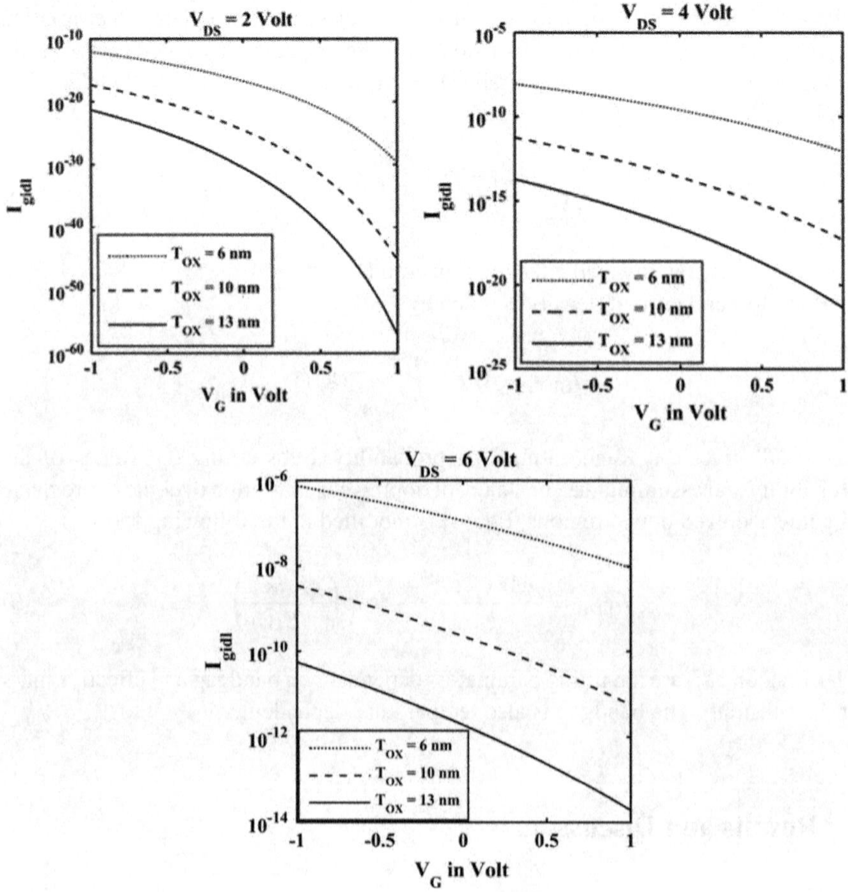

Fig. 1 Leakage current with gate voltage with different oxide thicknesses for **a** low V_{DS}, **b** moderate V_{DS}, **c** high V_{DS}

Computation is also carried with different doping concentration of semiconductor substrate, as shown in Fig. 2. It is found out that if doping is varied from moderate to high level, leakage current increases rapidly for a predefined bias set, as due to increase of thermionic current. But further heavy doping does not tune the magnitude sufficiently as the number of carriers responsible for leakage at the gate-drain end reaches saturation limit. Only, change in horizontal field can shift the level of current. Result is plotted for $V_{DS} = 4$ V and $V_{DS} = 6$ V.

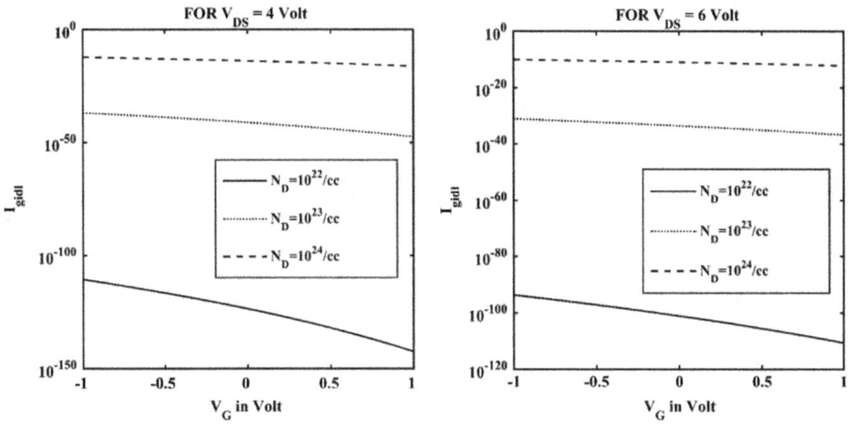

Fig. 2 Leakage current with gate voltage with different doping concentrations for **a** moderate V_{DS}, **b** high V_{DS}

Effect of high-*K* dielectric is analyzed in Fig. 3. From the plot, it is seen that with increasing oxide dielectric constant, leakage current increases. Since capacitance increases with increase in dielectric constant, charge storing capacity also increases, which in turn increases the leakage current. Actually high-*K* material enhances the magnitude of band bending. Comparative study is carried out with the conventional dielectric material for different magnitudes of V_{DS}. Thus, an optimization is required while use of high-*K* material in terms of gate-induced-drain-leakage.

Effect of gate overlap length is analyzed for different biasing conditions with high-*K* dielectric, represented in Fig. 4. In Fig. 4a, analysis is carried out for HfO_2, whereas in Fig. 4b, the same is made for TiO_2. In Fig. 4a, it is seen that reduction of overlapping region reduces the leakage current, as it reduces the amount of band bending. It may also be pointed out that for submicron range, the band-to-band tunneling current is comparatively large, which degrades the system performance as it creates large noise. In order to improve SNR, V_{DS} should be lowered; again it speaks for lower electric field, and henceforth, larger gate length. This is a compromising situation for nanoscale device design. When the result of Fig. 4a is compared with that of Fig. 4b, it is found out that increase of relative dielectric constant increases leakage current. This is due to the fact that high-*K* material bends the band more, and therefore, tunneling probability between filled valence band and empty conduction band increases.

Effect of temperature is studied next. From material science standpoint, higher temperature of the semiconductor increases bandgap, which, in turn, decreases the tunneling probability. Simulation is carried out for moderate and high values of V_{DS} separately, and findings are compared with the result obtained for SiO_2 (Fig. 5).

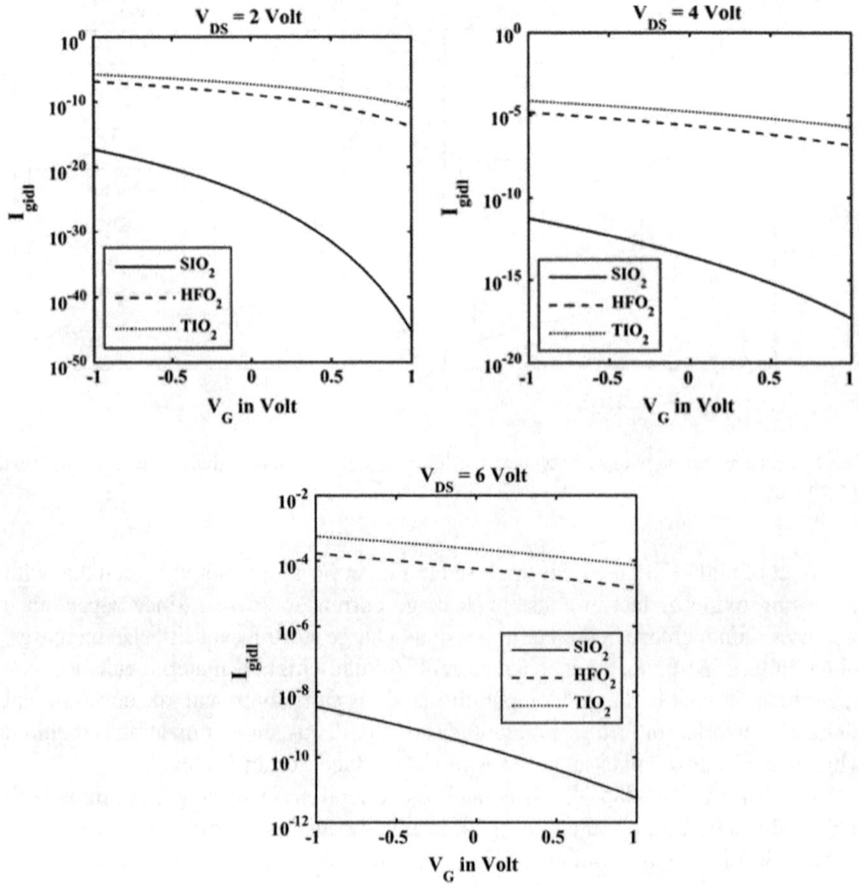

Fig. 3 Leakage current with gate voltage with different high-*K* materials for **a** low V_{DS}, **b** moderate V_{DS}, **c** high V_{DS}

4 Conclusion

Gate-induced-drain-leakage current is analytically investigated for moderate to high values of lateral electric field in presence of high-*K* dielectrics. Results are compared with that obtained for conventional SiO_2 material, and effect of gate overlap length is studied to control the leakage. Temperature variation is also incorporated as nanoscale device under high field may cause large Joule heat dissipation. Theoretical study verifies that substrate doping has an upper threshold limit for leakage current, over which leakage current becomes almost constant. Simulation is carried out for both negative and positive gate biases. Results are very critical for application of the device as memory.

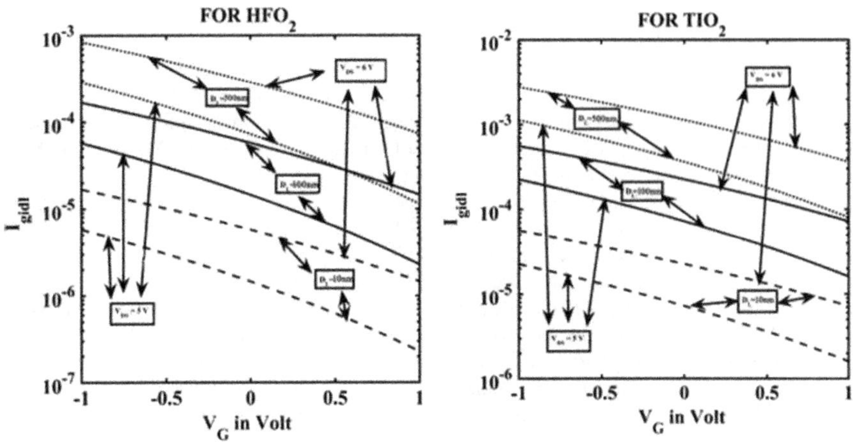

Fig. 4 Leakage current with gate voltage with different gate overlapped regions for **a** HfO_2, **b** TiO_2

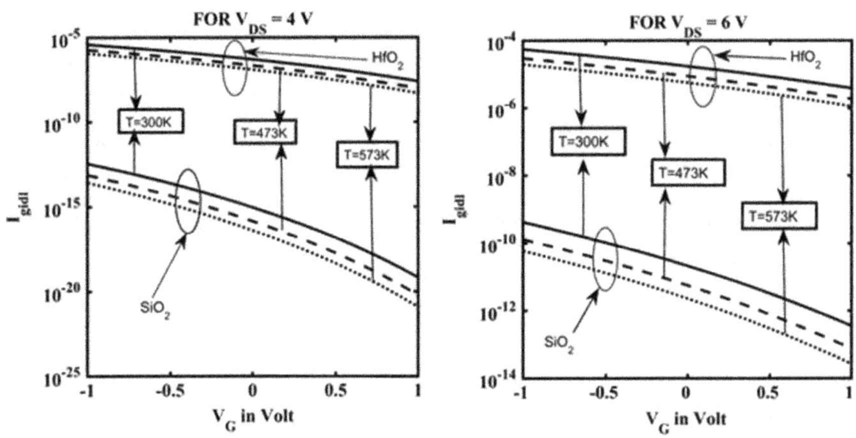

Fig. 5 Leakage current with gate voltage for HfO_2 and SiO_2 as dielectrics for **a** moderate V_{DS}, **b** high V_{DS}

References

1. ITRS Roadmap, 2007
2. Gupta, K.A., Anvekar, D.K., Venkateswarlu, V.: Comparative study and analysis of short channel effects for 180 nm and 45 nm transistors. In: Advances in Computing and Information Technology: Advances in Intelligent Systems and Computing, vol. 178, chap. 69, pp. 707–715 (2013)
3. Chaudhry, A., Kumar, M.J.: Controlling short-channel effects in deep-submicron SOI MOSFETs for improved reliability: a review. IEEE Trans. Device Mater. Rel. **4**(1), 99–109 (2004)
4. Cheng, H.W., Li, Y.: 16-nm multigate and multifin MOSFET device and SRAM circuits. In: International Symposium on Next-Generation Electronics (2010)

5. Subramanian, V.: Multiple gate field-effect transistors for future CMOS technologies. IETE Tech. Rev. **27**(6), 446–454 (2010)
6. Mudanai, S., Fan, Y.Y., Ouyang, Q., Tasch, A.F., Banerjee, S.K.: Modeling of direct tunneling current through gate dielectric stacks. IEEE Trans. Electron Devices **47**(10), 1851–1857 (2000)
7. Tiefeng, W., Zhichao, Z., Quan, W., Lizhi, G., Jing, L.: An algorithm on direct tunneling current model based on DIBL effect. In: International Conference on Chemical, Material and Food Engineering, Atlantis Press, pp. 719–722 (2015)
8. Amin, S.I., Sarin, R.K.: Direct tunneling gate current model for symmetric double gate junctionless transistor with SiO_2/High-K gate stacked dielectric. J. Semicond. **37**(3), 034001 (2016)
9. Oh, S.J., Yeow, Y.T.: A modification to the fowler-nordheim tunneling current calculation for thin MOS structures. Solid State Electron. **31**(6), 1113–1118 (1988)
10. Mondal, I., Dutta, A.K.: An analytical gate tunnelling current model for MOSFETs having ultrathin gate oxides. IEEE Trans. Electron Devices **55**(7), 1682–1692 (2008)
11. Hong, S.H., Jang, J.H., Park, T.J., Jeong, D.S., Kim, M., Hwang, C.S.: Improvement of the current-voltage characteristics of a tunneling dielectric by adopting a $Si_3N_4/SiO_2/Si_3N_4$ multilayer for flash memory application. Appl. Phys. Lett. **87**, 152106 (2005)
12. Cassan, E., Dollfus, P., Galdin, S., Hesto, P.: Calculation of direct tunneling gate current through ultra-thin oxide and oxide/nitride stacks in MOSFETs and H-MOSFETs. Microelectron. Reliab. **40**(4–5), 585–588 (2000)
13. Lai, P.T., Jingping, X., Liu, B.Y., Xu, Z.: New observation and improvement in GIDL of N-MOSFET's with various kinds of gate oxides under hot-carrier stress. In: IEEE International Conference on Semiconductor Electronics (1996)
14. Chang, L., Yang, K.J., Yeo, Y.C., Polishchuk, I., King, T.J., Hu, C.: Direct-tunneling gate leakage current in double-gate and ultrathin body MOSFETs. IEEE Trans. Electron Devices **49**(12), 2288–2295 (2002)
15. Lee, W.C., Hu, C.: Modeling CMOS tunneling currents through ultrathin gate oxide due to conduction- and valence-band electron and hole tunneling. IEEE Trans. Electron Devices **48**(7), 1366–1373 (2001)
16. Cai, J., Sah, C.T.: Gate tunneling currents in ultrathin oxide metal-oxide-silicon transistors. J. Appl. Phys. **89**(4), 2272–2285 (2001)
17. Nathan, V., Das, N.C.: Gate-induced drain leakage current in MOS devices. IEEE Trans. Electron Devices **40**(10), 1888–1890 (1993)
18. Chen, J., Chan, T.Y., Chen, I.C., Ko, P.K., Hu, C.: Subbreakdown drain leakage current in MOSFET. IEEE Electron Device Lett. **8**(11), 515–517 (1987)
19. Semenov, O., Pradzynski, A., Sachdev, M.: Impact of gate induced drain leakage on overall leakage of submicrometer CMOS VLSI circuits. IEEE Trans. Semicond. Manuf. **15**(1), 9–18 (2002)
20. Choi, Y.K., Ha, D., King, T.J., Bokor, J.: Investigation of gate-induced drain leakage (GIDL) current in thin body devices: single-gate ultra-thin body, symmetrical double-gate, and asymmetrical double-gate MOSFETs. Jpn. J. Appl. Phys. **42**, 2073–2076 (2003)
21. Dai, C.H., Chang, T.C., Chu, A.K., Kuo, Y.J., Ho, S.H., Hsieh, T.Y., Lo, W.H., Chen, C.E., Shih, J.M., Chung, W.L., Dai, B.S., Chen, H.M., Xia, G., Cheng, O., Huang, C.T.: Hot carrier effect on gate-induced drain leakage current in high-K/metal gate n-channel metal-oxide-semiconductor field-effect transistors. Appl. Phys. Lett. **99**, 012106 (2011)

Investigating Effect of Structural Parameters on Static Characteristics of Ultrathin DG MOSFET Using Taur's Model

Riya Chakraborty, Deepanwita Mondal and Arpan Deyasi

Abstract Drain current and pinch-off voltage of ultrathin double-gate MOSFET are analytically calculated based on Taur's model, where the centre potential is derived from Ortiz-Conde formulation. Drain current is computed for different structural parameters in lower nanometric range, and the effect of the high-*K* dielectric is investigated. Pinch-off voltage shift is therefore derived from the simulated findings and compared with the available findings followed by Ortiz-Conde. The result shows a measurable variation in the parameters, and the root cause is explained from the electrostatic point of view. Findings are important for conductance calculation.

Keywords Drain current · Taur's model · Pinch-off voltage · Structural parameters · Ortiz-Conde model · High-*K* dielectric

1 Introduction

Research on multiple-gate MOSFET gets attention in the last decade due to the severe constraint of short-channel effect [1] in low-dimensional devices and thereby requirement of precise gate control [2]. In submicron devices, more precisely when device dimension goes beyond 100 nm, the requirement of lower DIBL and moderate sub-threshold slope instigates several novel FET architectures, and double-gate MOSFET is one of the supreme candidates [3–5] among them. One branch of device engineering deals with tunnelling mechanism-based transistors, which results in single-electron transistor [6], tunnel field-effect transistor [7], etc., whereas another arena of research

R. Chakraborty · D. Mondal
Department of Electronic Science, A.P.C College, Barasat, India
e-mail: riyachakraborty167@gmail.com

D. Mondal
e-mail: mondaldeepanwita1996@gmail.com

A. Deyasi (✉)
Department of Electronics and Communication Engineering, RCC Institute of Information Technology, Kolkata, India
e-mail: deyasi_arpan@yahoo.co.in

J. K. Mandal et al. (eds.), *Information, Photonics and Communication*,
Lecture Notes in Networks and Systems 79,
https://doi.org/10.1007/978-981-32-9453-0_2

is gate engineering where multiple gates, as well as various architectures [8–10], are proposed for controlling electron transport. DG MOSFET is the result of later avenue of research as mentioned, and it offers excellent properties for analog [11] as well as digital [12] applications. Tied-gate architectures are preferred for higher current density [13], whereas independent-gate architecture offers lower threshold voltage [14], and henceforth preferred for low power design.

Inversion layer properties of DG MOSFET are analytically investigated by Palanichamy [15] after the simultaneous solution of Schrödinger and Poisson's equation, whereas weak inversion properties are computed by Bhartia [16] after inclusion of channel length modulation parameter. An explicit model was derived by Zhu et al. [17] following Taur's model, but that is only applicable for undoped structure. Hariharan [18] later included velocity saturation model for submicron device where gate length is considered 200 nm. The effect of the number of gates on drain current is investigated by Yu [19], followed by compact model development [20]. Very recently, Yu published [21] SPICE-compatible model for surface potential computation. In the present paper, drain current of symmetric DG MOSFET is analytically calculated based on Taur's model where centre potential is obtained from Ortiz-Conde analysis. The results are shown the closer agreement of data with published literature. Corresponding pinch-off voltage is calculated for different high-K dielectric and compared with that obtained for conventional SiO_2 material. The results are important for computing conductance of the device.

2 Mathematical Formulation

For long-channel DG MOSFET structure, the solution of 1D Poisson's equation gives [22]

$$\phi(z) = \phi_C - 2\phi_t \ln\left[\frac{t_{\text{sub}}}{2\beta}\sqrt{\frac{qn_i}{2\varepsilon_{\text{sub}}\phi_t}}\cos\left(\frac{2\beta z}{t_{\text{sub}}}\right)\right] \tag{1}$$

where the parameter is defined as [22]

$$\beta = \frac{t_{\text{sub}}}{2}\sqrt{\frac{qn_i}{2\varepsilon_{\text{sub}}\phi_t}}\exp\left[\frac{\phi_0 - \phi_C}{2\phi_t}\right] \tag{2}$$

Here t_{sub} defines the thickness of the substrate, ϕ_C is the quasi-Fermi potential for electrons inside the channel.

Drain current for the device is given by

$$I_{\text{DS}} = \mu_{n\text{eff}}\frac{W}{L}\frac{4\varepsilon_{\text{sub}}}{t_{\text{sub}}}(2\phi_t)2[f(\beta_s) - f(\beta_d)] \tag{3}$$

where

$$f(\beta) = \beta \tan \beta - 0.5\beta^2 + \frac{\varepsilon_{\text{sub}} t_{\text{ox}}}{\varepsilon_{\text{ox}} t_{\text{sub}}} \beta^2 \tan^2 \beta \tag{4}$$

In this case, ϕ_0 denotes the centre potential. In the present work, the value of centre potential is calculated following the Ortiz-Conde formulation [23].

Centre potential according to [23] is defined as

$$\phi_0 = U - \sqrt{U^2 - (V_{\text{GS}} - V_{fb})\phi_{0\text{max}}} \tag{5}$$

where 'U' and $\phi_{0\text{max}}$ are already defined.

In original Taur's model, centre potential is calculated from Eq. (1) with suitable boundary conditions, which is hereby replaced by Eq. (5).

3 Results and Discussion

Based on Eq. (3), we first calculated drain current for symmetric DG MOSFET, and the result is compared with that obtained from Ortiz-Conde model [23]. The result shows a very close agreement in saturation current, but a considerable difference in the active region. It is revealed from Fig. 1a that the slope of the active region is steeper in Ortiz-Conde model, whereas in the present paper, pinch-off voltage is delayed. This is due to the fact that the centre potential in the proposed model is a slowly varying function an affects the both source- and drain-end potentials, whereas in the model [23], the effect is overlooked. However, in the saturation region, the difference becomes negligibly small because of increasing drain voltage, which overcomes the effect of centre potential variation. The comparative study is also performed with the data obtained from Taur's model [24] and represented in Fig. 1b.

In Taur's model, centre potential is calculated directly from the function β [24]. Here that is computed from [23], and a noticeable difference is observed. This is due to the fitting of [23], where the function β, defined in Eq. (2), becomes a function of centre potential, and corresponding total potential function. High-K effect is investigated based on that modification. With the increase of dielectric constant, it is found that drain current decreases, as evident from Fig. 2. This is quite obvious, but another interesting fact that corresponding to the reduction of saturation current, pinch-off voltage takes a right shift.

The effect of dielectric thickness is investigated in Figs. 3 and 4 represent the substrate thickness effect. With the increase of dielectric thickness, current decreases and that is true for the substrate layer width also. Again pinch-off point shifts with the relative change and that is represented in tabular form.

The effect of back-gate voltage is calculated and plotted in Fig. 5. It is seen from the plot that the higher gate voltage leads to delay in pinch-off point due to the

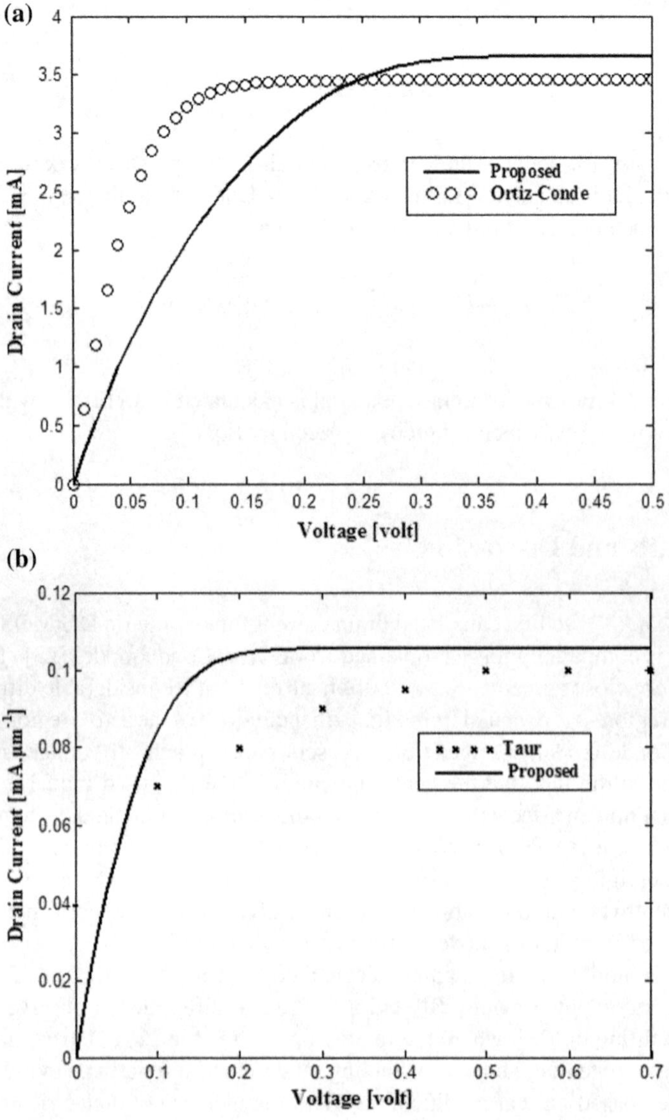

Fig. 1 **a** Comparative analysis of drain current with Ortiz-Conde model [23]. **b** Comparative analysis of drain current with Taur model [24]

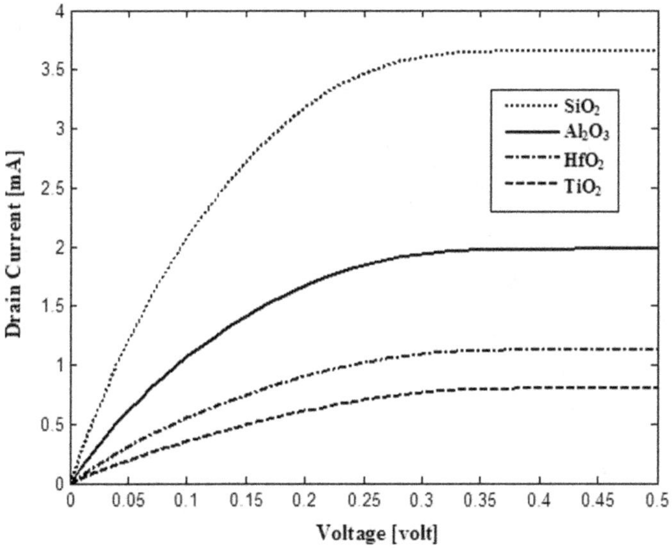

Fig. 2 Effect of high-*K* dielectric on drain current

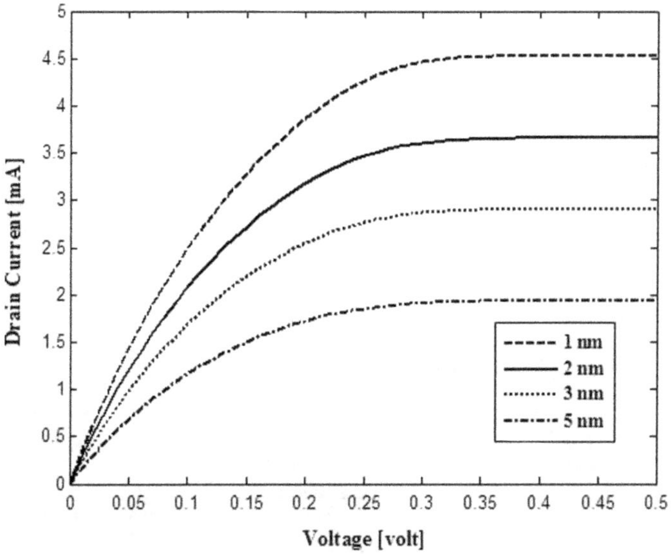

Fig. 3 Effect of dielectric thickness on drain current

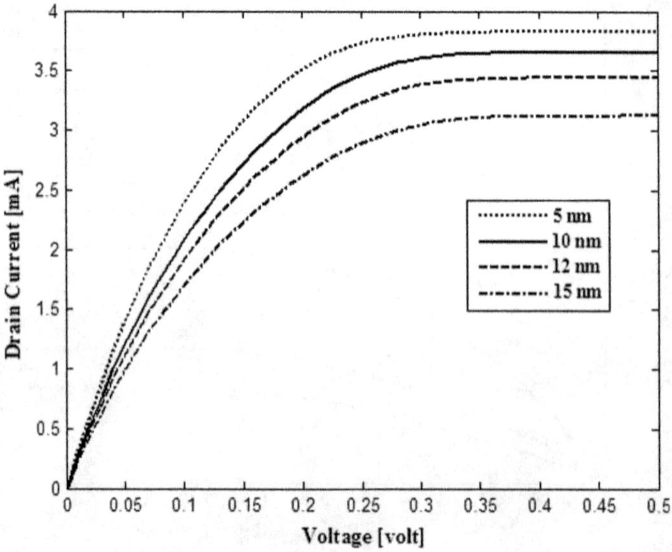

Fig. 4 Effect of substrate thickness on drain current

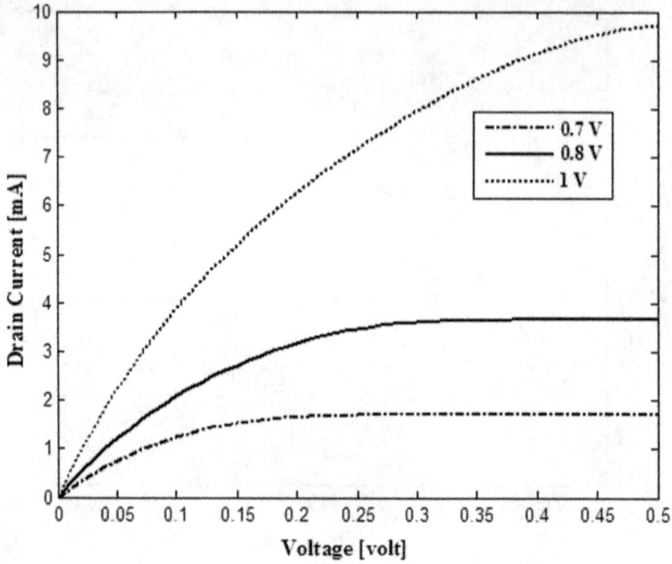

Fig. 5 Effect of back-gate voltage on drain current

Table 1 Pinch-off voltage for different dielectric thickness with two sets of back-gate voltage

t_{ox} (nm)	V_P (volt)							
	SiO_2		Al_2O_3		HfO_2		TiO_2	
	V_G = 0.8 V	V_G = 1 V	V_G = 0.8 V	V_G = 1 V	V_G = 0.8 V	V_G = 1 V	V_G = 0.8 V	V_G = 1 V
2	0.35	0.54	0.35	0.55	0.38	0.59	0.39	0.59
5	0.31	0.51	0.31	0.5	0.35	0.54	0.37	0.54
8	0.27	0.46	0.28	0.45	0.3	0.49	0.33	0.51
10	0.25	0.41	0.26	0.42	0.28	0.44	0.32	0.5

Table 2 Pinch-off voltage for different substrate thickness with two sets of back-gate voltage

t_{sub} (nm)	V_P (volt)							
	SiO_2		Al_2O_3		HfO_2		TiO_2	
	V_G = 0.8 V	V_G = 1 V	V_G = 0.8 V	V_G = 1 V	V_G = 0.8 V	V_G = 1 V	V_G = 0.8 V	V_G = 1 V
5	0.32	0.48	0.31	0.49	0.33	0.52	0.34	0.51
7	0.34	0.49	0.32	0.5	0.34	0.53	0.34	0.53
10	0.34	0.5	0.34	0.51	0.37	0.54	0.37	0.55
15	0.35	0.52	0.35	0.53	0.37	0.54	0.39	0.57

enhancement of the threshold barrier. But it also leads to higher saturation current due to DIBL factor. Corresponding data is shown in Tables 1 and 2.

4 Conclusion

Centre potential, as derived from Ortiz-Conde model, is put into the existing Taur's model, and both drain current and pinch-off voltages are computed from that. The results show a good agreement in the saturation region. The effect of structural parameters and back-gate voltage is calculated, and the shift of pinch-off voltage is vividly reflected from that results. Findings have greater significance for the computation of conductance characteristics.

References

1. Kim, Y.B.: Challenges for nanoscale MOSFETs and emerging nanoelectronics. In: Transactions on Electrical and Electronic Materials, vol. 11, no. 3, pp. 93–105 (2010)
2. Hiramoto, T., Nagumo, T.: Multi-gate MOSFETs with back-gate control. In: IEEE International Conference on IC Design and Technology (2006)

3. Yadav, V.K., Rana, A.K.: Performance analysis of double gate MOSFETs with different gate dielectric. In: IEEE International Conference on Signal Processing, Computing and Control (2012)
4. Xiong, D.X., Sun, L., Yan, L.X., Qi, H.R.: A comparative study of double gate MOSFET with asymmetric barrier heights at source/drain and the symmetric DG-SBFET. In: International Workshop on Junction Technology (2009)
5. Dasgupta, A., Das, R., Chakraborty, S., Dutta, A., Kundu, A., Sarkar, C.K.: Comparisons between dual and tri material gate on a 32 nm double gate MOSFET. Nano **11**(10), 1650117 (2016)
6. Willy, F., Darma, Y.: Modeling and simulation of single electron transistor with master equation approach. J. Phys: Conf. Ser. **739**, 012048 (2016)
7. Yang, Z.: Tunnel field-effect transistor with an L-shaped gate. IEEE Electron Device Lett. **37**(7), 839–842 (2016)
8. Singh, M., Kumar, G., Bordoloi, S., Trivedi, G.: A study on modeling and simulation of multiple-gate MOSFETs. J. Phys: Conf. Ser. **759**, 012093 (2016)
9. Singh, D., Panda, S., Mohapatra, S.K., Pradhan, K.P., Sahu, P.K.: Static performance analysis on UTB-SG and DG MOSFETs with Si and III–V channel materials. In: International Conference on High Performance Computing and Applications (2014)
10. Jimenez, D., Iniguez, B., Sune, J., Marsal, L.F., Pallares, J., Roig, J., Flores, D.: Continuous analytic I-V model for surrounding-gate MOSFETs. IEEE Electron Device Lett. **25**(8), 571–573 (2004)
11. Zhang, W., Fossum, J.G., Mathew, L., Du, Y.: Physical insights regarding design and performance of independent-gate FinFETs. IEEE Trans. Electron Devices **52**(10), 2198–2206 (2005)
12. Keane, J., Eom, H., Kim, T.H., Sapatnekar, S., Kim, C.: Stack sizing for optimal current drivability in subthreshold circuits. IEEE Trans. Very Large Scale Integr. Syst. **16**(5), 598–602 (2008)
13. Woo, H.J., Jin, K.C., Kyu, C.Y.: Universal potential model in tied and separated double-gate MOSFETs with consideration of symmetric and asymmetric structure. IEEE Trans. Electron Devices **55**(6), 1472–1479 (2008)
14. Liu, Y.X., Masahara, M., Ishii, K., Tsutsumi, T., Sekigawa, T., Takashima, H., Yamauchi, H., Suzuki, E.: Flexible threshold voltage FinFETs with independent double gates and an ideal rectangular cross-section Si-Fin channel. In: IEEE International Electron Devices Meeting (2003)
15. Palanichamy, V., Balamurugan, N.B.: Analytical modeling of drain current, capacitance and transconductance in symmetric double-gate MOSFETs considering quantum effects. Int. J. Nanosci. **12**(1), 135005 (2013)
16. Bhartia, M., Chatterjee, A.K.: Modeling the drain current and its equation parameters for lightly doped symmetrical double-gate MOSFETs. J. Semicond. **36**(4), 044003 (2016)
17. Zhu, Z., Zhou, X., Rustagi, S.C., See, G.H., Lin, S., Zhu, G., Wei, C., Zhang, J.: Analytic and explicit current model of undoped double-gate MOSFETs. Electron. Lett. **43**(25), 1–2 (2007)
18. Hariharan, V., Vasi, J., Rao, V.R.: Drain current model including velocity saturation for symmetric double-gate MOSFETs. IEEE Trans. Electron Devices **55**(8), 2173–2180 (2008)
19. Yu, B., Song, J., Yuan, Y., Lu, W.Y., Taur, Y.: A unified analytic drain-current model for multiple-gate MOSFETs. IEEE Trans. Electron Devices **55**(8), 2157–2163 (2008)
20. Karatsori, T.A., Tsormpatzoglou, A., Theodorou, C.G., Ioannidis, E.G., Haendler, S., Planes, N., Ghibaudo, G., Dimitriadis, C.A.: Development of analytical compact drain current model for 28 nm FDSOI MOSFETs. In: 4th International Conference on Modern Circuits and Systems Technologies, pp. 1–4 (2015)
21. Yu, F., Huang, G., Lin, W., Xu, C.: An analytical drain current model for symmetric double-gate MOSFETs. AIP Adv. **8**, 045125 (2018)
22. Taur, Y., Liang, X., Wang, W., Lu, H.: Analytic drain current model for DG MOSFETs. IEEE Electron Device Lett. **21**(5), 245–247 (2004)

23. Ortiz-Conde, A., Garcia-Sanchez, F.J., Muci, J.: Analytical solution for drain current of undoped symmetric dual-gate MOSFETs. Solid State Electron. **49**(4), 640–647 (2005)
24. Taur, Y., Liang, X., Wang, W., Lu, H.: A continuous, analytic drain-current model for DG MOSFETs. IEEE Electron Device Lett. **25**(2), 107–109 (2004)

Implementation of L-Shaped Dielectric Double Metal Dual-Gate TFET Toward Improved Performance Characteristics and Reduced Ambipolarity

Bijoy Goswami, Sutanni Bhowmick, Arindam Haldar, Goutika Paul, Debadipta Basak and Subir Kumar Sarkar

Abstract In this work, we have proposed an L-shaped Dielectric Double Metal Dual-Gate TFET, which has been experimentally demonstrated and investigated for the first time. This structure offers several benefits over the conventional planar TFET in terms of reduced OFF-state current and hence suppressed ambipolar conduction. The source region has been extended in vertical direction for better tunneling at the source-channel interface. The material for the L-shaped oxide is varied and accordingly three structural variations have been simulated to obtain the optimized performance characteristics: (a) L-shaped HfO_2-SiO_2 Double Metal Dual-Gate TFET (LS-HS-DM-DG TFET: vertical portion has HfO_2 and horizontal portion has SiO_2), (b) L-shaped SiO_2-HfO_2 Double Metal Dual-Gate TFET (LS-SH-DM-DG TFET: vertical portion has SiO_2 and horizontal portion has HfO_2), and (c) L-shaped SiO_2-SiO_2 Double Metal Dual-Gate TFET (LS-SS-DM-DG TFET: total L-shape comprised of SiO_2). All the fundamental device parameters have been analyzed to demonstrate the tunneling phenomena in the L-shaped Dielectric gate-engineered TFET structure under various biasing conditions. All the simulations have been performed in 2D using Silvaco, Atlas.

B. Goswami (✉) · S. Bhowmick · A. Haldar · G. Paul · D. Basak · S. K. Sarkar
Department of ETCE, Jadavpur University, Kolkata 700032, India
e-mail: bijoy.ete@aec.ac.in

S. Bhowmick
e-mail: sutannibhow94@gmail.com

A. Haldar
e-mail: arindamhaldar169@gmail.com

G. Paul
e-mail: munaigoutika2@gmail.com

D. Basak
e-mail: db.rini1991@gmail.com

S. K. Sarkar
e-mail: su_sircir@yahoo.co.in

J. K. Mandal et al. (eds.), *Information, Photonics and Communication*,
Lecture Notes in Networks and Systems 79,
https://doi.org/10.1007/978-981-32-9453-0_3

Keywords L-shaped dielectric double metal dual-gate TFET · LS-HS-DM-DG TFET · LS-SH-DM-DG TFET · LS-SS-DM-DG TFET · Ambipolar conduction suppression

1 Introduction

Conventional MOSFET structures have several drawbacks in the nanoscale range which have been compensated by TFET devices. The fundamental phenomenon upon which TFET is based is quantum mechanical band-to-band tunneling (BTBT), as opposed to carrier transport by thermionic injection in MOSFETs. The lowest subthreshold slope (SS) that can be achieved in a MOSFET is 60 mV/decade at room temperature. TFETs have demonstrated SS lower than this value owing to the difference in operating principles. Also steeper switching behavior in TFETs makes this device suitable for high-speed and low-power applications. Additionally, TFETs are immune to short-channel effects and exhibit negligibly low leakage currents in the range of femto amperes (fA), making them suitable for high-frequency applications.

Despite the inherent merits of TFETs, the primary area of concern in these devices is ambipolar conduction, owing to band-to-band tunneling at the drain-channel junction. Also the ON current is considerably low as compared to MOSFETs. To suppress the ambipolar conduction, various structures have been previously demonstrated. These include gate–drain underlap structure, [1–5] which significantly reduces the ambipolar conduction, although current driving capacity is compromised in the process and other approaches [6–9]. In this work, we have opted for a gate-engineered L-shaped Dielectric Double Metal Dual Gate TFET to suppress ambipolar conduction. The gate engineering is obtained by using different materials along the L-shaped dielectric region in the double metal dual-gate structure. The electric field profile has been studied to show reverse band bending in the drain-channel interface. This has the effect of greatly reducing the tunneling probability at the drain end of the structure, such that ambipolar conduction may be effectively suppressed. The OFF current is in the range of 10^{-18} A, thereby offering high I_{ON}/I_{OFF} and hence reduced subthreshold slope. The structures have been named according to the corresponding dielectric materials used in the gate-oxide regions, respectively, (a) L-shaped HfO_2-SiO_2 Double Metal Dual-Gate TFET (LS-HS-DM-DG TFET: vertical portion has HfO_2 and horizontal portion has SiO_2), (b) L-shaped SiO_2-HfO_2 Double Metal Dual-Gate TFET (LS-SH-DM-DG TFET: vertical portion has SiO_2 and horizontal portion has HfO_2), and (c) L-shaped SiO_2-SiO_2 Double Metal Dual-Gate TFET (LS-SS-DM-DG TFET: total L-shape comprised of SiO_2). The effects of varying the gate bias and oxide permittivities on the band diagram and other parameters such as the electric field and potential profile have been analyzed, and the ramifications of the structural variations and materials used for optimal evaluation have been reported.

2 Device Structure and Simulation Setup

The 2D cross-sectional view of the proposed n-channel TFET structure is shown in Fig. 1. (a) L-shaped HfO_2-SiO_2 Double Metal Dual-Gate TFET (LS-HS-DM-DG TFET), (b) L-shaped SiO_2-HfO_2 Double Metal Dual-Gate TFET (LS-SH-DM-DG TFET), and (c) L-shaped SiO_2-SiO_2 Double Metal Dual-Gate TFET (LS-SS-DM-DG TFET). The total length of the device is 44 nm while the channel length is 14 nm. The device has L-shaped oxide region in both upper and lower side. By varying the components of the oxide layers, three different structures are constructed. For LS-SS-DM-DG TFET, the total oxide layer is of SiO_2, while the other two structures contain combination of SiO_2 and HfO_2. The Source region is extended with vertical dimension of 25 nm with 10 nm horizontal length. The drain regions in all the three devices are of heavily doped N + type with doping concentration of 10^{18} cm^{-3} with vertical length of 14 nm and horizontal extent of 10 nm. The doping concentration of the source is 10^{20} cm^{-3} (p-type) and that of the channel region is 10^{12} cm^{-3} (p-type). For symmetry, air is considered as the dielectric above and below the drain region. Thickness of the oxide region t_{ox} above and below the channel is 2 nm whereas the vertical section of the oxide has thickness of 3 nm. The double metal comprises of metals M1 and M2 with work functions 5.1 and 4.1 eV, respectively. All the simulations have been performed using Silvaco Atlas version 5.20.2.R by including appropriate models such as nonlocal BTBT (BBT.NONLOCAL), Auger recombination model, Lombardi mobility model (CVT), the Shockley–Read–Hall, Fermi–Dirac statistics, bandgap narrowing model, and Drift-diffusion carrier transport have been used. The band structures have been demonstrated under various biasing conditions.

3 Results and Discussions

3.1 Band Diagram Analysis

Figure 2 compares the ON-state band diagram of the (a) L-shaped HfO_2-SiO_2 Double Metal Dual-Gate TFET (LS-HS-DM-DG TFET), (b) L-shaped SiO_2-HfO_2 Double Metal Dual-Gate TFET (LS-SH-DM-DG TFET),and (c) L-shaped SiO_2-SiO_2 Double Metal Dual-Gate TFET (LS-SS-DM-DG TFET). The band structures suggest better tunneling profile, reduced leakage, and lower ambipolar conduction in comparison to conventional planar TFETs. The tunneling length, which is inversely related to the tunneling probability, is lower for LS-HS-DM-DG TFET and LS-SH-DM-DG TFET as compared to LS-SS-DM-DG TFET. This indicates better tunneling in the former two structures. Reverse band bending, as observed from the energy band diagrams, is indicative of reduced ambipolar conduction. There is a stronger reverse band bending at the channel-drain interface of LS-HS-DM-DG TFET and LS-SH-DM-DG TFET as compared to LS-SS-DM-DG TFET. So, it can be concluded that LS-HS-DM-DG TFET and LS-SH-DM-DG TFET structures are more immune to ambipolar

Fig. 1 2D cross-sectional structure for **a** L-shaped HfO_2-SiO_2 double metal dual gate TFET (LS-HS-DM-DG TFET), **b** L-shaped SiO_2-HfO_2 double metal dual gate TFET (LS-SH-DM-DG TFET), and **c** L-shaped SiO_2-SiO_2 double metal dual gate TFET (LS-SS-DM-DG TFET)

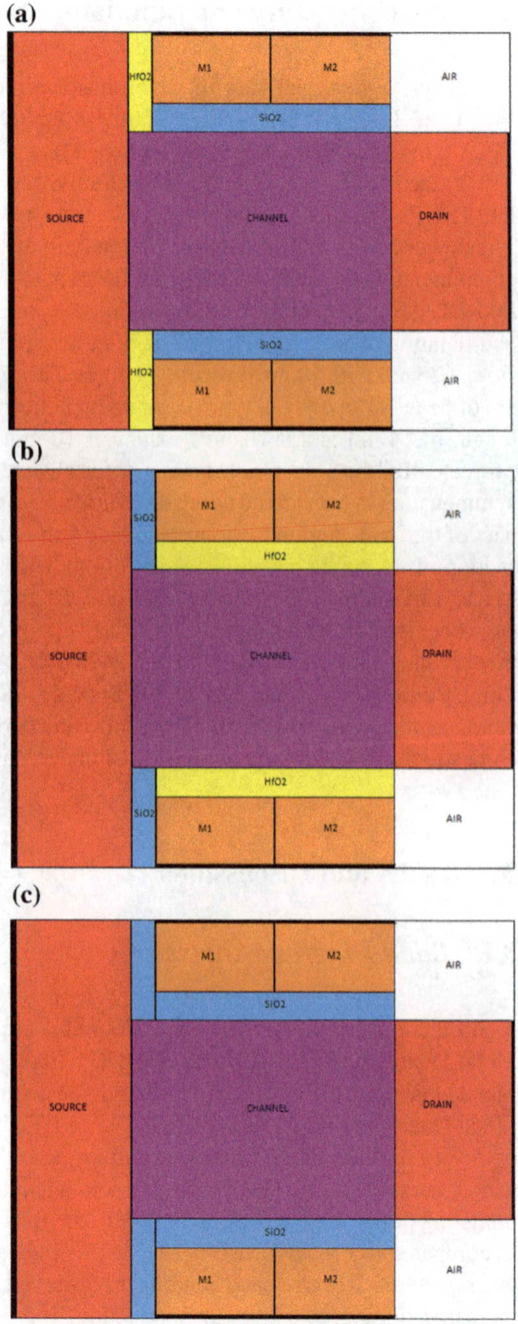

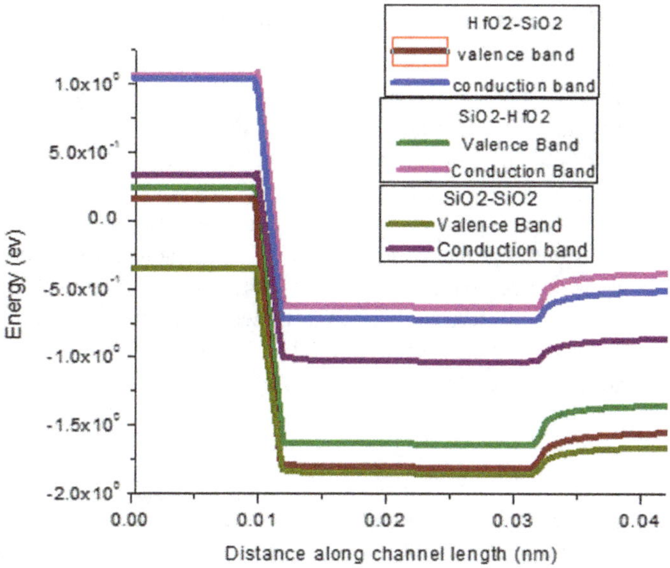

Fig. 2 ON-state band diagram of **a** L-shaped HfO_2-SiO_2 Double Metal Dual-Gate TFET (LS-HS-DM-DG TFET), **b** L-shaped SiO_2-HfO_2 Double Metal Dual-Gate TFET (LS-SH-DM-DG TFET), and **c** L-shaped SiO_2-SiO_2 Double Metal Dual-Gate TFET (LS-SS-DM-DG TFET)

conduction as compared to the LS-SS-DM-DG TFET structure. The energy band diagram can be further optimized by modulation in dimensions of the structure and variation in the doping profile.

Figure 3 shows conduction band energy for the LS-HS-DM-DG TFET structure. The conduction band energy profile of the other two structures is almost similar to the one in Fig. 3. As expected, the conduction band energy is highest at the source end, reduces at the source-channel interface, and is lowest at the drain end.

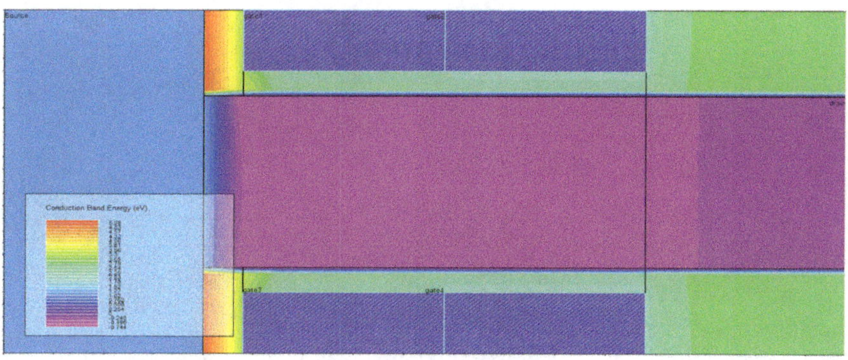

Fig. 3 Conduction band energy profile for LS-HS-DM-DG TFET

3.2 Potential Profile and Electric Field Profile Analysis

A comparative analysis of the potential profile is presented in Fig. 4 for the three structures. The peak in the potential profile at the source-channel interface is indicative of the overall tunneling. It can be seen from the figure that overall tunneling is better for LS-HS-DM-DG TFET as compared to LS-SH-DM-DG TFET and LS-SS-DM-DG. The potential profile of the latter two structures is almost overlapped. Higher value of potential at the channel-drain interface indicates that the electrons which tunnel at the source-channel interface can effectively reach the drain.

The electric field for all three structures is shown in Fig. 5. Maximum overshoot at the source-channel interface is seen for LS-HS-DM-DG TFET suggesting highest degree of tunneling in this structure. The peak in the electric field profile at the

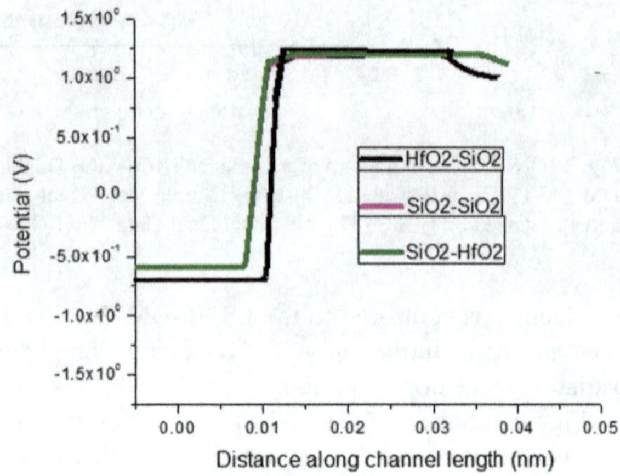

Fig. 4 Potential profile of **a** L-shaped HfO_2-SiO_2 Double Metal Dual-Gate TFET (LS-HS-DM-DG TFET), **b** L-shaped SiO_2-HfO_2 Double Metal Dual-Gate TFET (LS-SH-DM-DG TFET), and **c** L-shaped SiO_2-SiO_2 Double Metal Dual-Gate TFET (LS-SS-DM-DG TFET)

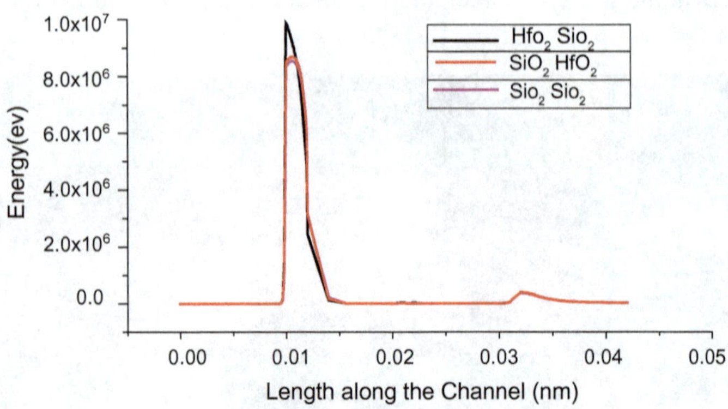

Fig. 5 Electric field profile of **a** L-shaped HfO_2-SiO_2 Double Metal Dual-Gate TFET (LS-HS-DM-DG TFET), **b** L-shaped SiO_2-HfO_2 Double Metal Dual-Gate TFET (LS-SH-DM-DG TFET), and **c** L-shaped SiO_2-SiO_2 Double Metal Dual-Gate TFET (LS-SS-DM-DG TFET)

channel-drain interface acts as a barrier to reverse tunneling. This peak indicates immunity to DIBL. The peak of the electric field profile at the channel-drain interface is lower than that in the source-channel interface implying reduced hot carrier effect at the drain end. The band-to-band tunneling profile is shown for all the three structures in Fig. 6. The sharpest reverse peak at the source-channel interface for the LS-HS-DM-DG TFET structure is indicative of superior band-to-band tunneling for this structure. This is also reflected in the transfer characteristics of the structures.

LS-HS-DM-DG TFET shows most favorable outcomes in terms of potential profile and electric field profile. The contour diagrams of the potential profile and the electric field profile for the LS-HS-DM-DG TFET structure have been shown in Figs. 7 and 8, respectively.

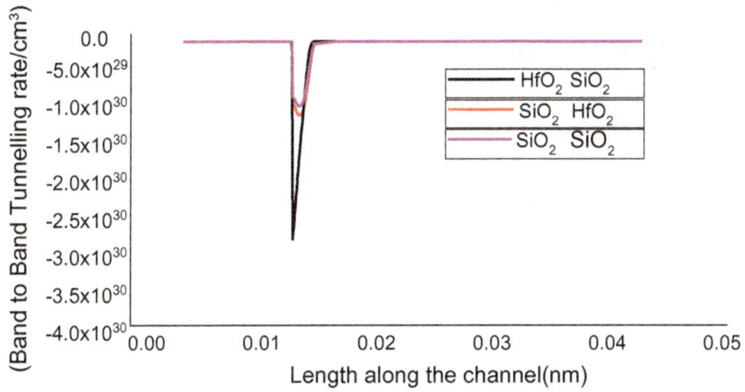

Fig. 6 Band-to-Band tunneling profile of **a** L-shaped HfO_2-SiO_2 Double Metal Dual-Gate TFET (LS-HS-DM-DG TFET), **b** L-shaped SiO_2-HfO_2 Double Metal Dual-Gate TFET (LS-SH-DM-DG TFET), and **c** L-shaped SiO_2-SiO_2 Double Metal Dual-Gate TFET (LS-SS-DM-DG TFET)

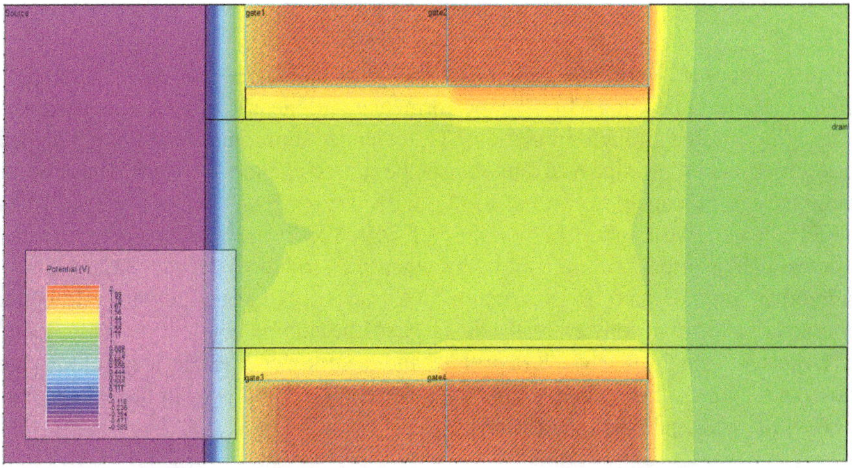

Fig. 7 Potential profile of LS-HS-DM-DG TFET

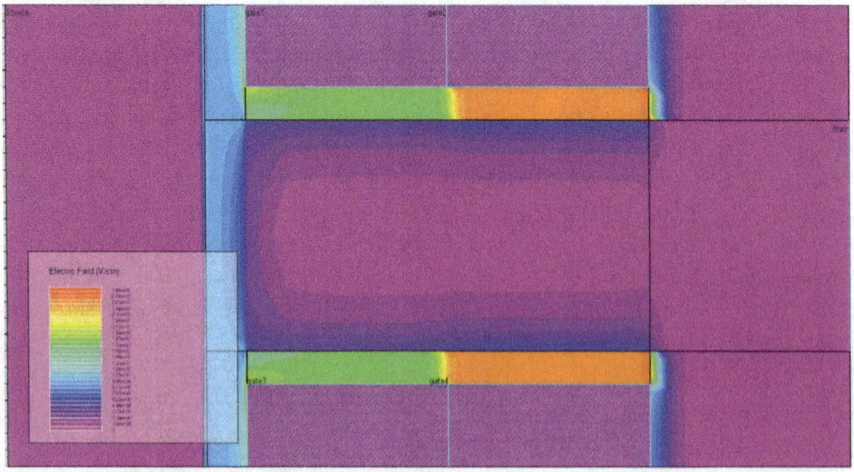

Fig. 8 Electric field profile of LS-HS-DM-DG TFET

3.3 *Transfer Characteristics*

The transfer characteristics are shown for all the three structures in Fig. 9. Highest I_{ON}/I_{OFF} ratio is observed for LS-HS-DM-DG TFET. The subthreshold slope is 24 mV/dec for LS-HS-DM-DG TFET, 34 mV/dec for LS-SH-DM-DG TFET, and 60 mV/dec for LS-SS-DM-DG TFET. Also, suppressed ambipolar conduction is clearly observed from the transfer characteristics of LS-HS-DM-DG TFET.

4 Conclusion

In this proposed work, the effects of the gate-engineered L-shaped Dielectric Double Metal Dual-Gate TFET on performance parameters have been experimentally demonstrated through simulations. The extended source along with the L-shaped dielectric ensures the improved tunneling at the source-channel interface. Three structures have been named according to the dielectric material used. LS-HS-DM-DG TFET shows lowest subthreshold slope of 24 mV/dec and superior potential and electric field profiles as compared to the other two structures. Hence, it can be considered as the most optimized structure in this work. All structures indicate better characterization than conventional TFETs. Also, the ambipolar conduction is reduced as indicated from the electric field profiles and the transfer characteristics of the proposed structures. This outcome can be further explored and used in the field of low power, RF, and analogue applications.

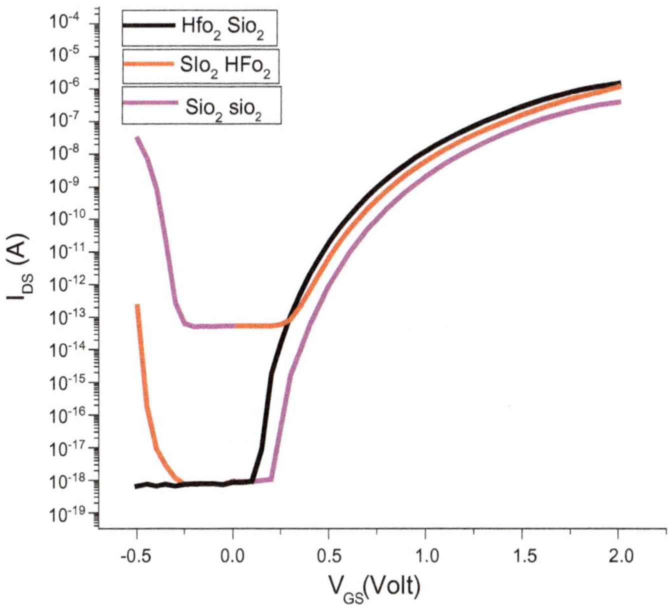

Fig. 9 Transfer characteristics of **a** L-shaped HfO_2-SiO_2 Double Metal Dual-Gate TFET (LS-HS-DM-DG TFET), **b** L-shaped SiO_2-HfO_2 Double Metal Dual-Gate TFET (LS-SH-DM-DG TFET), and **c** L-shaped SiO_2-SiO_2 Double Metal Dual-Gate TFET (LS-SS-DM-DG TFET)

References

1. Bagga, N., Sarkhel, S., Sarkar, S.K.: Recent research trends in gate engineered tunnel FET for improved current behavior by subduing the ambipolar effects: a review. In: 2015 International Conference on IEEE Computing, Communication & Automation (ICCCA), pp. 1264–1267 (2015)
2. Khatami, Y., Banerjee, K.: Steep subthreshold slope n-and p-type tunnel-FET devices for low-power and energy-efficient digital circuits. IEEE Trans. Electron Devices **56**, 2752–2761 (2009)
3. Saurabh, S., Kumar, M.J.: Novel attributes of a dual material gate nano scale tunnel field-effect transistor. IEEE Trans. Electron Devices **58**(2), 404–410 (2011)
4. Bagga, N., Sarkar, S.K.: An analytical model for tunnel barrier modulation in triple metal double gate TFET. IEEE Trans. Electron Devices **62**(7), 2136–2142
5. Yadav, S., Sharma, D., Chandan, B.V., Aslam, Md, Soni, D., Sharma, N.: A novel hetero-material gate-underlap electrically doped TFET for improving DC/RF and ambipolar behaviour. Superlattices Microst. **117**, 9–17 (2018). Elsevier
6. Lattanzio, L., De Michielis, L., Ionescu, A.M.: Complementary germanium electron–hole bilayer tunnel FET for sub-0.5-V operation. IEEE Electron Device Lett. **33**(2), 167–169 (2012)
7. Ghosh, B., Akram, M.W.: Junctionless tunnel field effect transistor. IEEE Electron Device Lett. **34**(4), 584–586 (2013)
8. Wang, L., Yu, E., Taur, Y., Asbeck, P.: Design of tunneling field effect transistors based on staggered hetero junctions for ultra low power applications. IEEE Electron Device Lett. **31**(5), 431–433 (2010)
9. Goswami, B., Bhattachariee, D., Dash, K.D., Bhattacharya, A., Sarkar, S.K.: Demonstration of T-shaped channel tunnel field-effect transistors. In: (IEMENTech) 2018 2nd International Conference on Electronics, Materials Engineering & Nano-Technology

Memristor-Based Circuit Design Approaches and Future Challenges

Ramesh Kumar and Surajit Mandal

Abstract Memristor is a two-terminal element which remembers its most recent memristance or history, and is characterized by its nonlinear relation between charge and magnetic flux and hysteresis loop in the current versus voltage plane. Memristor which is investigated as the fourth basic passive element the other three are resistor, capacitor, and inductor. With the increase of research interest in memristor, it is promising potential in developing the future memory devices. Here, we review the memristor science and connected technology and discuss different aspects of the memristor. The memristor circuit design using its equivalent PSPICE memristor model is presented and time diagrams are obtained. Memristors behavior with other passive element like resistor and its interrelation is investigated. Here, we cover principles of memristors systems, useful models of memristors, practical memristors circuit implementations, and highlighted prospective utilization of memristors using the passive circuit elements like resistor. The related characteristics and the future challenges of memristors are discussed in this communication.

Keywords TiO_2 memristor · Memristor modeling · PSPICE · Memristor applications

1 Introduction

In 1971, Leon Chua suggested the existence of memristor which is a basic circuit element like other passive elements [1]. Chua extended the traditional relationship between voltages, current, charge, and magnetic flux. The memristor theory was further termed to memristive devices in 1976 [2]. In 2008, Hewlett-Packard (HP) Laboratory fabricated the first memristor and published results that described the memristor device [3]. Considering the switching capability, a memristor-switch is

R. Kumar (✉) · S. Mandal
B.P. Poddar Institute of Management and Technology, Kolkata 700052, India
e-mail: ramesh.nestor@gmail.com

S. Mandal
e-mail: surajitmandal@yahoo.co.in

J. K. Mandal et al. (eds.), *Information, Photonics and Communication*,
Lecture Notes in Networks and Systems 79,
https://doi.org/10.1007/978-981-32-9453-0_4

connected at the intersection of every row line and column line. When the switch between row and column is closed, row line gets electrically connected to column line. By applying voltage to the ends of the terminals, the idea is to open and close the switches. Open switch is indicated by 0 and closed switch is indicated by 1. In crossbar switch, it consists of rows and columns which can be platinum lines and a 40 nm cube TiO_2 is sandwiched between the layers and the junction to create memristor [4]. As the memristor is hypothetical nonlinear resistive device which perceives memory characteristics, Chua named this element "Memristor" in 1971. The memristor has varying resistance which is also called memristance. The unit of memristance is same as that of resistance which is expressed in Ohm (Ω). For the memristor-based circuit design, analyze and simulate different models like linear ion-drift model, nonlinear ion-drift model, Simmons Tunnel Barrier model, threshold adaptive memristor model, and macro model have been proposed [5].

2 Application Perspective of Memristors

2.1 Memristor for Memories

It is convinced that memristor will change the memory circuit design in the future. Memristors can store data indefinitely, as they can remember their state [6], whereas in conventional DRAM the capacitor needs energy to retain the information. The interesting properties of memristor would be a significant role for circuit designing and tasks computation. Similar to atomic switch, the entity and control variables of memristor are atoms and charge whereas its state and output variables are charge and electron [7]. To improve overall energy, low bias and low power consumption to make suitable for fast switching the memristor performance will be significance [8]. The ability of memristor to remember the resistance is favorable in the artificial neural network and its behavior seems as that of synapse, which mimics aspects of human brain for investigating and expecting [9].

2.2 Breakthrough in Memristor

A non-volatile memory lies between low power and high density whereas memristor mimics biological synapse of neural networks. Memory and processor coexist in novel computer architecture. Memristor is going to lead a dominant role in the recognition of inventive neuromorphic computing architecture merging memory and computational abilities. This fundamental step exploring high to expand as the memristor tends to bring data close to computation capabilities [10].

3 Practical Memristors

3.1 Equations of HP's Memristor

It is difficult to obtain the exact relation for HP's memristor due to ion drift mechanisms and coupling complexity. Figure 1 shows the construction of TiO_2-based memristor and the memristance $M(q)$ and.their relationship based on the assumptions by HP [11] is given in Eq. (4).

$$M(t) = R(w) = R_{\text{ON}}\frac{w(t)}{D} + R_{\text{OFF}}\left(1 - \frac{w(t)}{D}\right) \tag{1}$$

$$\frac{\mathrm{d}w(t)}{\mathrm{d}x} = \mu_v \frac{R_{\text{ON}}}{D} i(t) \tag{2}$$

$$w(t) = \mu_v \frac{R_{\text{ON}}}{D} q(t) + w_{\text{o}} \tag{3}$$

Equation (3) is the integral form of Eq. (1). Here, $q(t)$ represents the integral of the current to time. The memristance $M(q)$ can be obtained from Eqs. (1) and (2) given by this

$$M(q) = R_{\text{OFF}} - \frac{R_{\text{OFF}} - R_{\text{ON}}}{D} w(t) \tag{4}$$

Here, D represents the width of the doped and undoped layer, and μ_v the dopant mobility. HP memristor model proposed by HP used the TiO_2-based devices; the equivalent circuit of the proposed memristor structure is mentioned in Fig. 1.

Memristors are compatible with conventional CMOS devices concerning manufacturing and operating voltage. It is an analog circuit which behaves digitally, and the resistance of the memristor typically represents a binary value [12].

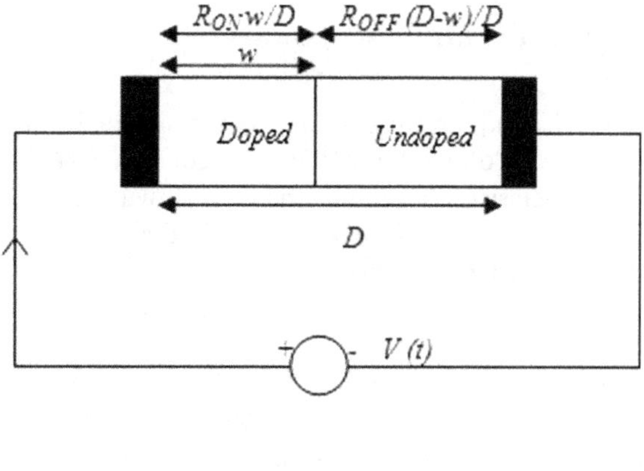

Fig. 1 Construction of the HP TiO_2 memristor

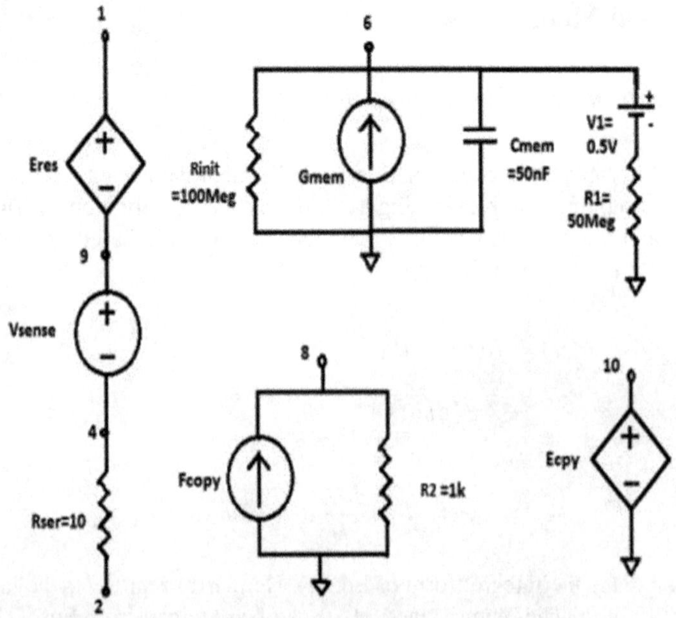

Fig. 2 Electrical model of memristor PSPICE simulation [13]

3.2 *Memristor Circuit*

The memristor model used by using SPICE in this discussion is similar to that of the Biolek model shown in Fig. 2. We obtained the simulation results by using the input sine voltage of amplitude 1.0 V and frequency 500 Hz. The different curves obtained by using the referred model are shown in the result and discussion section. Thus, the model can be assumed to be verified and work properly.

4 Results and Discussions

Input and output voltage of memristor is plotted with respect to time. Here, the input sine voltage of amplitude 1 V and frequency 500 Hz was taken. The output voltage of the memristor has a small phase shift with respect to the input voltage due to the memristance effect which is shown in Fig. 3. For the software simulation, we have used the Cadence environment.

The memristor current and voltage waveforms for the memristor model are obtained in Fig. 4.

In Fig. 5, we use memristor with same polarity and obtained the corresponding voltages in Fig. 6. Here, the resistance R1 = 2 kΩ and R2 = 1 kΩ is used. The input and output voltages waveforms at memristor M1 and M2 terminals are plotted in

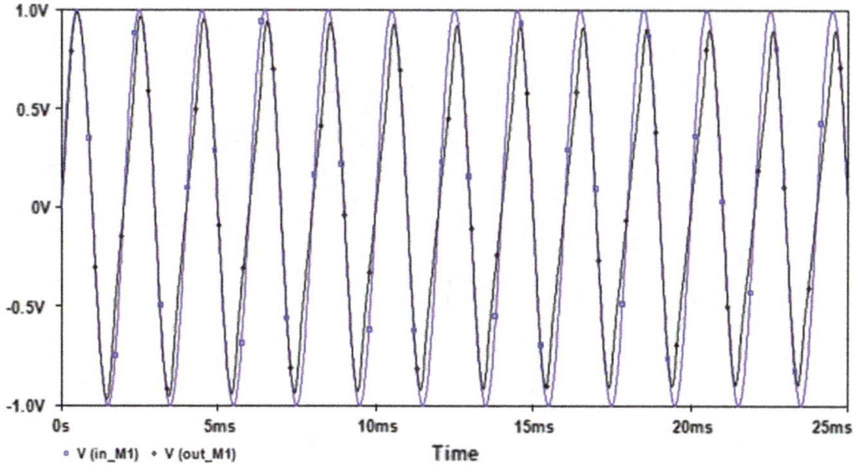

Fig. 3 Time diagram of input and output waveform for the memristor M1

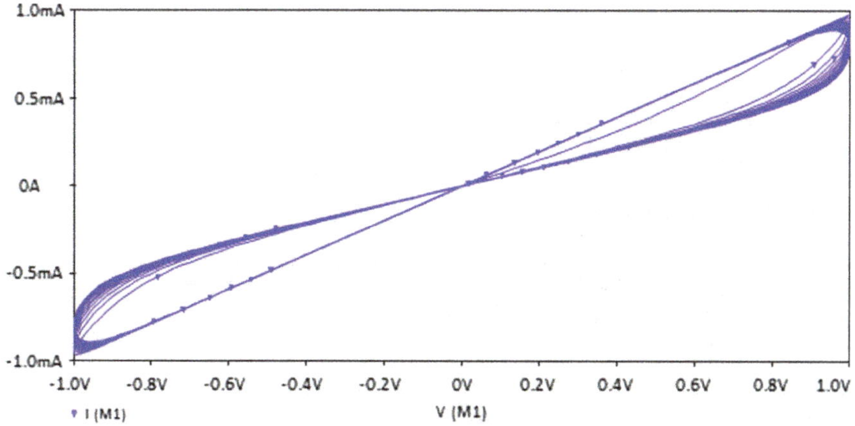

Fig. 4 Current and voltage hysteresis

Fig. 6, and resistance across M1, R1 and voltage in Fig. 7, respectively. The voltages *V* (in_M2) and *V* (out_M2) are almost similar in nature.

Figure 8 shows the circuit when the memristor polarity is changed, and the input and output voltage waveform with their timing diagram is obtained in Figs. 9 and 10, respectively.

Two memristor M1 and M2 are also connected in parallel in Fig. 11a and b and the diagrams are shown in Figs. 12 and 13, respectively.

Here, the hysteresis curve is obtained, Fig. 12 whereas more linearity is observed in Fig. 13 when the 1 K resistor is connected in series with M1 and M2, respectively. The equivalent memristance can be obtained by the ratio between applied voltage

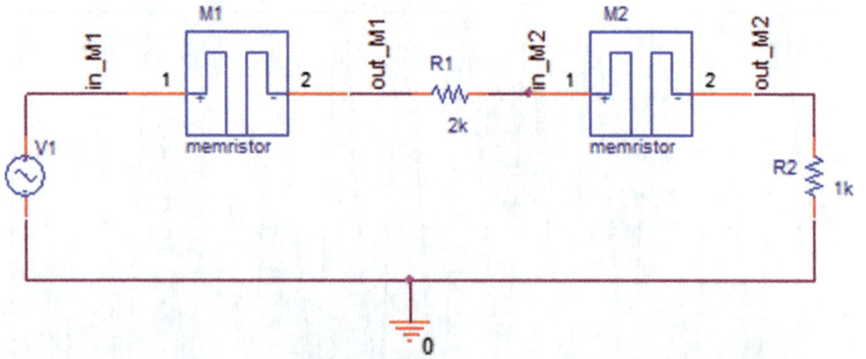

Fig. 5 Memristor same polarity connection

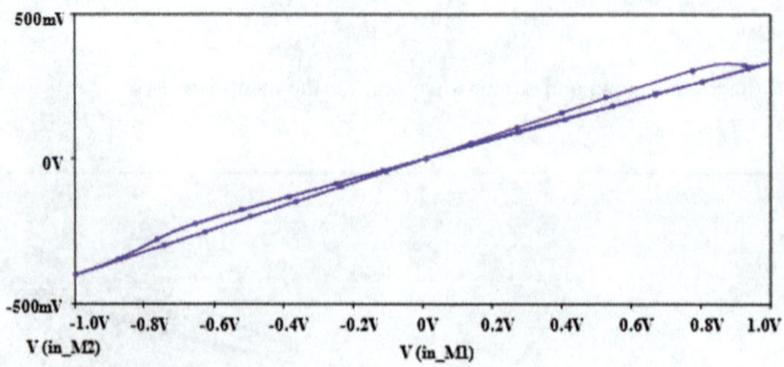

Fig. 6 M1 output and input voltages curve

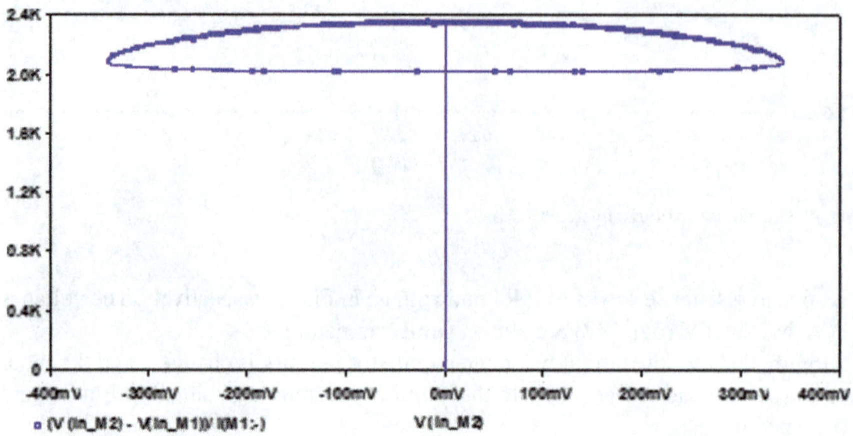

Fig. 7 Resistance across M1, R1 and voltage

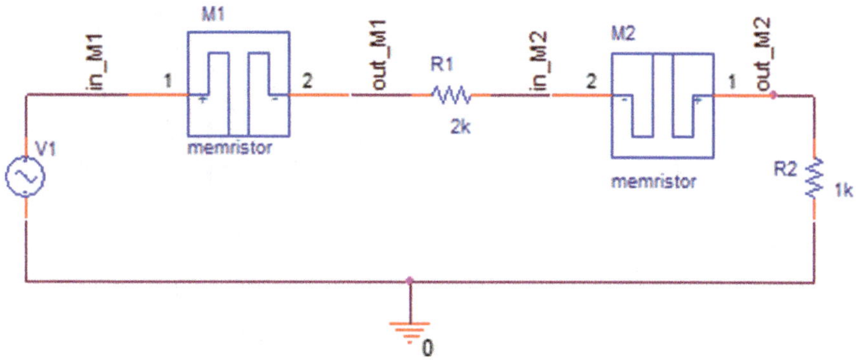

Fig. 8 Memristor opposite polarity connection

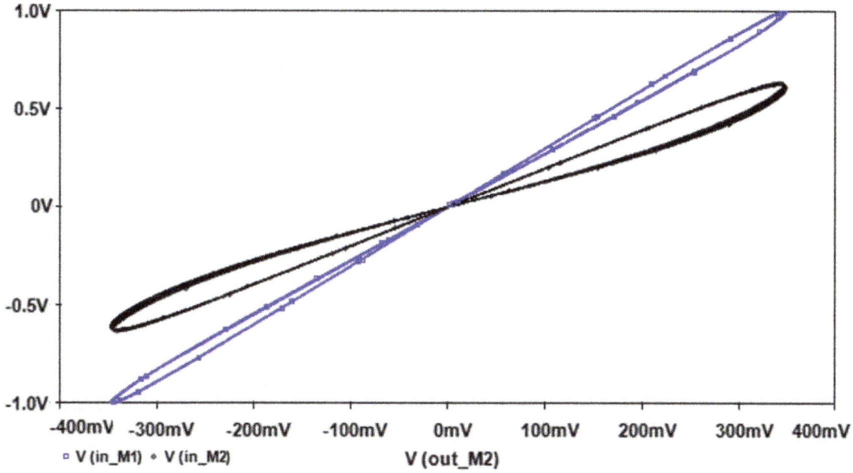

Fig. 9 Output and input voltages at M2, M1

and the current flowing through memristors. Under dc biasing, for the higher values of resistance a constant current is found in the circuit whereas for the lower values of resistance a significant value of current is obtained (Fig. 14). Figure 15 has ohmic nature due to considering the resistance.

From the input and output waveforms for the memristor M1 and M2 shown in Figs. 6 and 13, it have been found the similarities between them in terms of current and voltage. Particularly, analyzing the behavior of memristor connected in series and parallel in circuits is challenging because of its polarity. However, the device characteristics of memristors as compared to resistor are not exactly similar even though their units are same. Also dc circuit model of memristors should be further tested to study the memristor characteristics. For understanding the memristors resistor effect, more investigation is needed and it should be tested with other passive

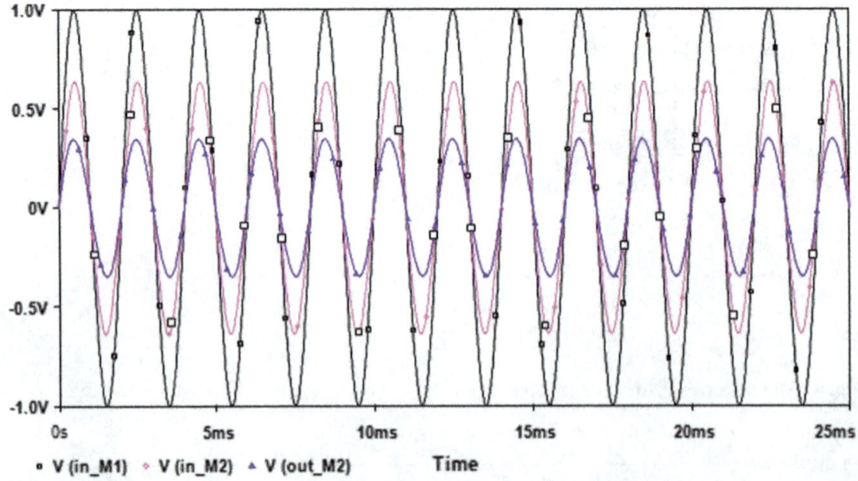

Fig. 10 Time diagram of input and output voltage of memristor M1, M2

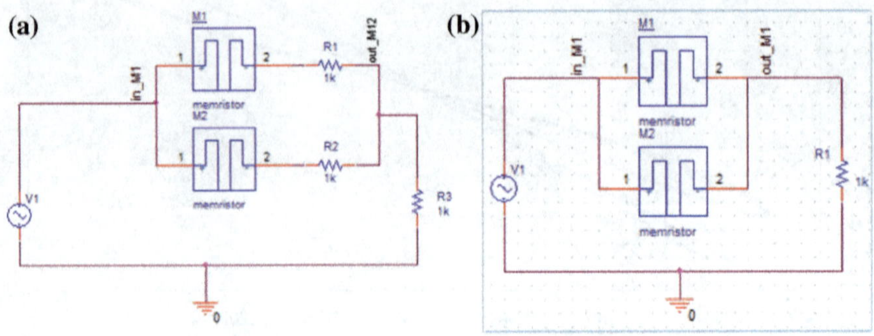

Fig. 11 **a** Memristor connection with resistor. **b** Memristor parallel connection

elements like capacitor and inductors to place it in passive element category. As the PSPICE models enables the modification to study the nonlinearity behavior of memristive elements. However PSPICE allows negative values for the passive elements but my cause instabilities during the analysis.

5 Conclusion

Memristors find applications in analog circuits such as chaos circuit, oscillator, amplifier, neural network etc. They are also found to perform logical operations in digital gates, and reconfigurable logic gates [14]. Memristor emulators using OP-AMPs and optocoupler are also implemented and tested [15]. The designs that were proposed

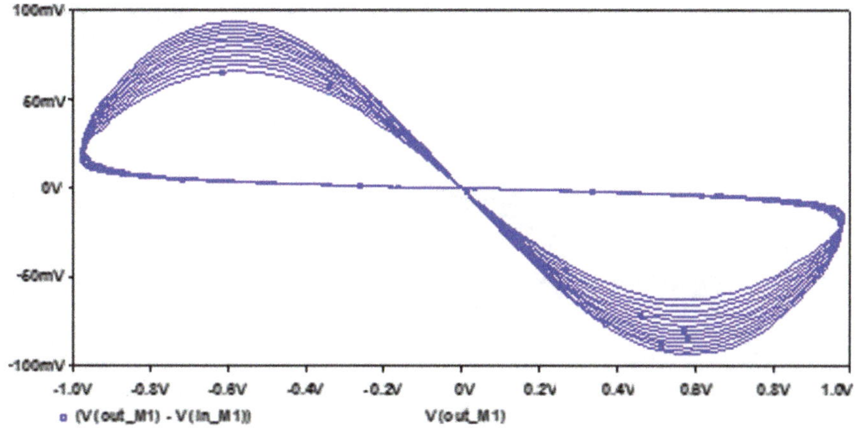

Fig. 12 Output voltages at M1

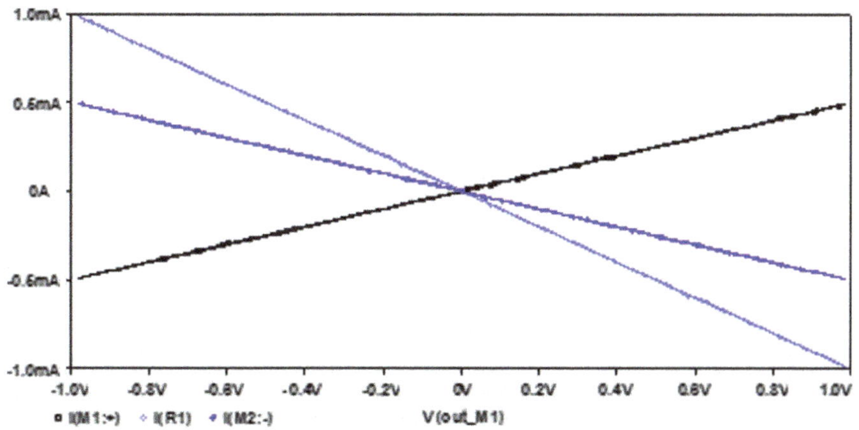

Fig. 13 Current at memristors, resistor and output voltage

for the designing and further the layout for memristor are not yet capable for accurate simulation of its working. We need better designs, also there are no libraries available for physical designing of memristor-based circuit, it still is a concept that needs more research. Many purposes have been accomplished till now as a result of deep research like the uses of memristor at analog circuit level, crossbar memory design, digital implementation of memristor is still not well researched and it needs more research to accomplish digital implementation because of the fact that Boolean logic cannot be applied directly. Many researches continue to put it as a physically feasible circuit device. For the validation of different modification made by the researchers, the relevant information should be provided by the manufacture. Much research work has been published concerning the memristive devices, yet these groups are waiting

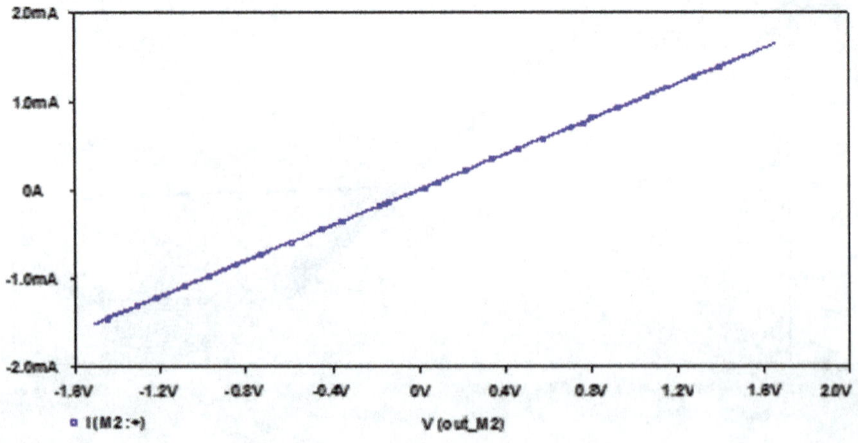

Fig. 14 Current (M2:+) versus voltage (out_M2) plot

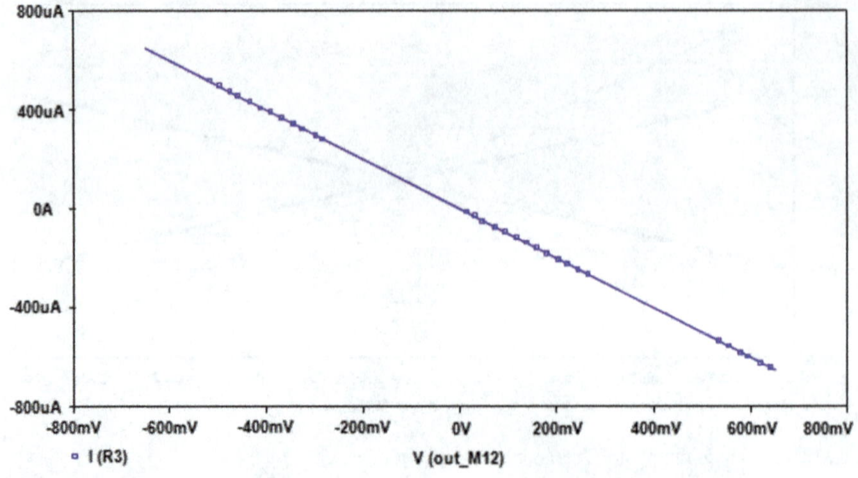

Fig. 15 Current (R3) versus voltage (out_M12) plot

for the physical availability of these elements in an integrated circuits form only a few products are available for learning capability only. Till now memristors are not commercially (except a few) available which should be available such that practical exposure and testing, learning ability, performance, speed, and other issues can be tested and compared with available memory elements. There are still many challenges ahead to be solved, such as those related to CMOS-memristor mixed system integration.

References

1. Chua, L.: Memristor—the missing circuit element. IEEE Trans. Circ. Theory **18**(5), 507–519 (1971)
2. Chua, L., Kang, S.M.: Memristive devices and systems. Proc. IEEE **64**, 209–223 (1976)
3. Strukov, D.B., Snider, G.S., Stewart, D.R., Williams, R.S.: The missing memristor found. Nature **453**, 80–83 (2008)
4. Vourkas, I., Sirakoulis, G.Ch.: A novel design and modeling paradigm for memristor-based crossbar circuits. IEEE. Trans. Nanotechnol. **11**(6) (2012)
5. Keshmiri, V.: A study of the memristor models and applications. Linköping Univ., Linköping, Sweden, Tech. Rep. LITH-EX-11/4455 (2014)
6. Dongale, T.D.: An overview of fourth fundamental circuit element-'the memristor'. Available at: https://nanohub.org/resources/16590
7. Prodromakis, T., Toumazou, C.: A review on memristive devices and applications. In: 2010 17th IEEE International Conference on Electronics, Circuits, and Systems (ICECS), pp. 936–939. IEEE (2010)
8. Pino, R.E., Campbell, K.A.: Compact method for modeling and simulation of memristor devices. In: Proceeding of International Symposium on Nanoscale Architecture, pp. 1–4 (2010)
9. Xiao, S., Xie, X., Wen, S., Zeng, Z., Huang, T., Jiang, J.: GST-memristor-based online learning neural networks. Neurocomput. J. **272**(C), 677–682 (2018)
10. Joglekar, Y.N., Wolf, S.J.: The elusive memristor: properties of basic electrical circuits arXiv:0807.3994 v2 [cond-mat.meshall], pp. 1–24, 13 Jan 2009
11. Radwan, A.G., Zidan, M.A., Salama, K.N.: HP memristor mathematical model for periodic signals and DC. In: Proceedings IEEE. International Midwest Symposium on Circuits System (MWSCAS), pp. 861–864 (2010)
12. Dahl, S.G., Ivans, R.: Modeling memristor radiation interaction events and the effect on neuromorphic learning circuits. In: Proceedings of the International Conference on Neuromorphic Systems, Article No. 1, Knoxville, TN, USA, pp. 23–26 (2018)
13. Biolek, Z., Biolek, D., Biolkova, V.: SPICE model of memristor with nonlinear dopant drift. Radioeng. J. **18**(2), 210–214 (2009)
14. Radwan, M.E.F.A.G.: Chapter 2 Memristor: Models Types and Applications (2015)
15. Chilukuri, M., Jung, S.: A high frequency memristor emulator circuit. In: IEEE Dallas Circuits Systems Conference (DCAS), pp. 1–4 (2015)

Communication System and Network

Performance Analysis of AntHocNet Based on NS2

Khondekar Lutful Hassan and Jyotsna Kumar Mandal

Abstract In this paper a simulation-based performance analysis of AntHocNet has been performed. Ant colony optimization (ACO)-based routing protocol, AntHocNet and Ad hoc On-Demand Distance Vector (AODV) routing protocol have been considered as MANET routing protocols. For the simulation purpose, NS2 (NS2.34) has been taken as simulation tools, and two attributes are taken for analysis purpose. The number of nodes (20, 40, 60, 80 and 100) and percentage of node mobility (20, 40, 60 and 80%) are used for the purpose of analysis. The analysis has been done under two protocol systems, AODV and AntHocNet to compute the comparative characteristics of both protocols. No of generated packets, throughput, average and end-to-end (e2e) delay are considered as the parameters for the purpose of comparisons and performance analysis. It is seen from the analysis that performance of AntHocNet is better than AODV in respect to all parameters which are considered for the analysis.

Keywords MANET · AntHocNet · AODV · NS2 · ACO · Bio-inspired routing

1 Introduction

Mobile ad hoc network (MANET) is defined as a set of discreet mobile nodes formed a discreet network. It is infrastureless, and mobility is the basic characteristic of MANET. Any nodes can enter into the network as well as can leave the network at any time. As mobility is the basic characteristic of the network, routing is very challenging job in such dynamic network. In the conventional MANET, routing protocol is classified broadly in two types, one is proactive routing protocol and

K. L. Hassan (✉)
Department of Computer Science and Engineering, Aliah University, Kolkata, West Bengal, India
e-mail: klhassan@yahoo.com

J. K. Mandal
Department of Computer Science and Engineering, University of Kalyani, Kalyani, Nadia, West Bengal, India
e-mail: jkm.cse@gmail.com

J. K. Mandal et al. (eds.), *Information, Photonics and Communication*, Lecture Notes in Networks and Systems 79,
https://doi.org/10.1007/978-981-32-9453-0_5

another is reactive routing protocol. In this protocol, a routing table is created before the data transmission in the network. Each node continuously maintains the updates of its routing table entire the network. That's why overhead is very high in this type of routing protocol. DSDV is the example of table-driven routing protocol. Reactive routing protocol creates the route when it is required to transmit the data, that's why this type of routing protocol is called on demand. Example of reactive routing protocol is AD hoc On-Demand Distance Vector (AODV) and Dynamic Source Routing (DSR).

Nowadays, researchers are trying to develop routing protocol in nonconventional way using various types of optimization technique algorithm [1–3], like bio-inspired algorithm. AntHocNet is the example of routing protocol using bio-inspired optimization technique, named Ant colony optimization (ACO) [4–7]. AntHocNet is also a kind of hybrid multipath routing protocol. ACO is the part of swarms intelligent by which ant can solve a complex problem using cooperation. ACO is a kind of multi-agent metaheuristic searching algorithm based on food searching of real ants in the real world. When ant leaves from the nest for searching the food at that time, it spreads pheromones as land mark in the route by which it travels. Using the land mark of pheromones, new ant in real life can find the shortest path from food to home. In the AntHocNet routing protocol will spread the pheromones agent for the route. It can create multiple routes.

Simulation environment is described in Sect. 2 of this paper and comparison of the performance is described in Sect. 3, conclusion is drawn in Sect. 4.

2 Simulation Environment

For simulation purpose, network simulator 2 (ns 2.34) is taken which is a Linux based and widely used in research and industry. For the implementation of AntHocNet in NS2, few routing protocol files are inherited and edited to arrange for the proper environment. Few existing configuration files in NS2 have also been edited for the required environment. For the purpose of simulation and analysis, few parameters are considered as fixed parameters which are given in Fig. 1, and few parameters are considered as variable parameters which are given in Table 1. Fixed parameters are those parameters which are considered as constant for every scenario as well as variable parameters are those which are changed for every scenario for generating comparisons of the result.

The simulation is performed in 1000×1000 m^2 area. Comparison of the performance is measured in terms of generated packet, throughput and average e2e delay.

Speed of the nodes	100 m/s
Time of simulation	100 sec
Mobility model	Random way point
Channel Type	Channel/Wireless Channel
Radio-propagation model	Propagation/TwoRayGround
Network Interface Type	Phy/WirelessPhy
MAC type	Mac/802_11
Interface queue type	CMUPreQueue
Link Layer Type	LL
Antenna Model	Antenna/Omniantenna
Max packet in ifq	50

Fig. 1 Fixed parameter for the simulations for each scenario

Table 1 Variable parameter for the simulation for each scenario

Routing protocols	AODV and AntHocNet
No of nodes	20, 40, 60, 80 and 100
% of node mobility	20, 40, 60 and 80%

3 Result and Analysis

For the purpose of the analysis, three attributes are taken. These are generated packets, throughput and average end-to-end delay. Various node configurations are used against various node mobilities. Section 3.1 deals with the issue of generated packets. Section 3.2 has been analyzed the throughput of under various mobility of nodes along with different number of nodes. Section 3.3 deals with average end-to-end delay.

3.1 Generated Packet

This parameter is measured as total number of packets generated in every experiment by all nodes. Figures 2, 3, 4 and 5 show the comparison of generated packets when 20, 40, 60 and 80% nodes are moving, respectively.

From Fig. 2, it is seen that when 20% nodes are moving at that time AntHocNet generates more packets than AODV routing protocol in every scenario.

When 40% nodes are moving at that time the result is same as 20% nodes are moving. Here, AntHocNet generates more packets than AODV routing in every scenario like Fig. 3.

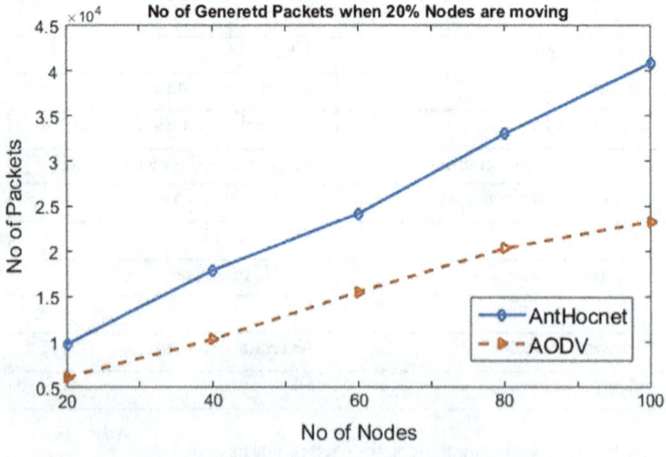

Fig. 2 Number of generated packet when 20% nodes are moving

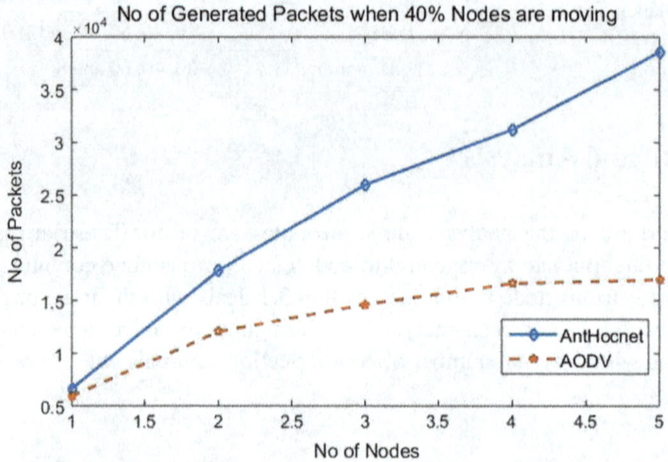

Fig. 3 Number of generated packet when 40% nodes are moving

When 60% nodes are moving at that time the Fig. 4 shows a little abnormal at node number is 80, the overall number of generated packets of AntHocNet decreased, but at that time generated packet is still more than number of packets generated by AODV.

When 80% nodes are moving, the total generated packet of AntHocNet is more than AODV routing protocol. It is observed in the Fig. 5 that when node number increased the generated packets also increased gradually.

From Figs. 2, 3, 4 and 5, it is seen that performance of generating packets by AntHocNet is always greater than AODV routing protocols in every scenario.

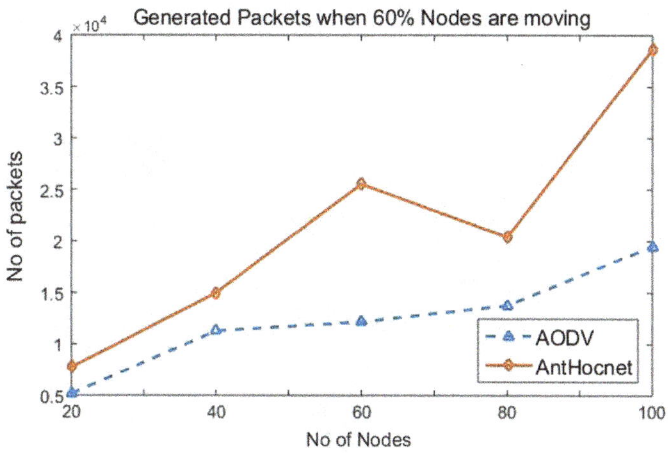

Fig. 4 Number of generated packet when 60% nodes are moving

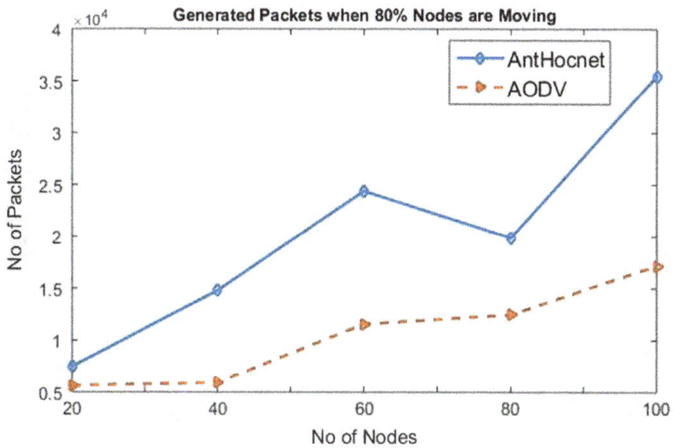

Fig. 5 Number of generated packet when 80% nodes are moving

3.2 Throughput

Throughput is measured as rate of data transferred in the entire network. Figures 6, 7, 8 and 9 show the comparison of throughput when 20, 40, 60 and 80% nodes are moving, respectively.

When 20% nodes are moving at that time throughput of AntHocNet is greater than the throughput of AODV. But at node number 20, the throughput of AODV is little greater.

When 40% nodes are moving at that time throughput of AntHocNet is higher than the throughput of AODV routing protocol.

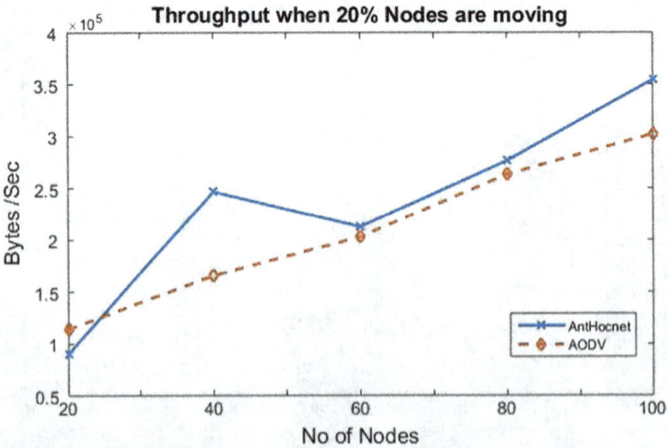

Fig. 6 Throughput when 20% nodes are moving

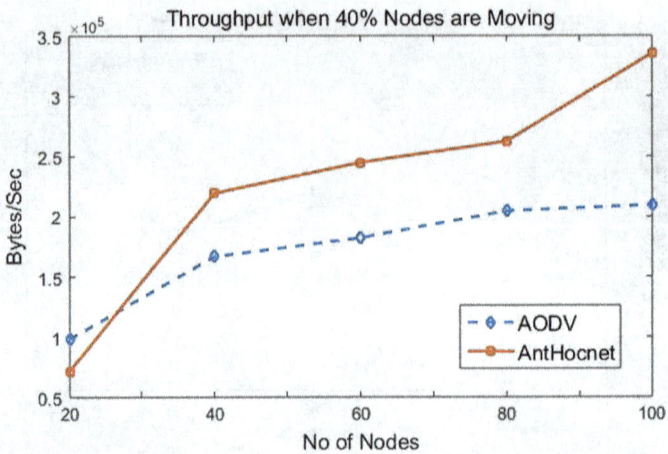

Fig. 7 Throughput when 40% nodes are moving

When 60% nodes are moving at that time the behavior of throughput graph are more and less same like the throughput when 40% nodes are moving. But at the node number between 20 and 40, the throughput of AODV is greater than the throughput of AntHocNet. But rest of the cases, the throughput of AntHocNet is greater than throughput of AODV.

When 80% nodes are moving at that time behavior of the throughput of AntHocNet is slightly abnormal. When node number is lower (20) at that time throughput of AntHocNet is less than the throughput of AODV. When node number is 40–100 at that time throughput of AntHocNet is higher than AODV routing protocol. But when

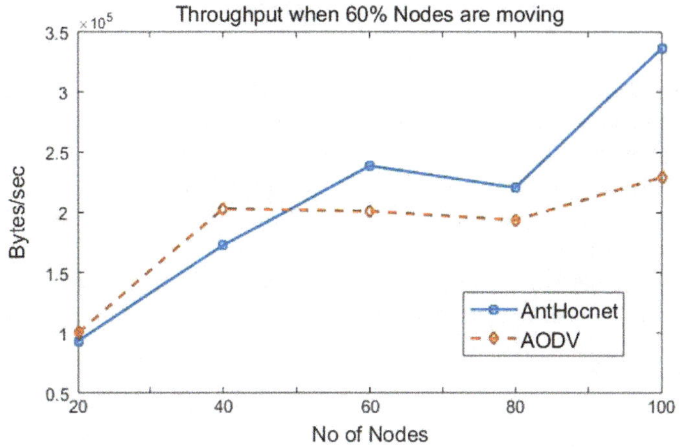

Fig. 8 Throughput when 60% nodes are moving

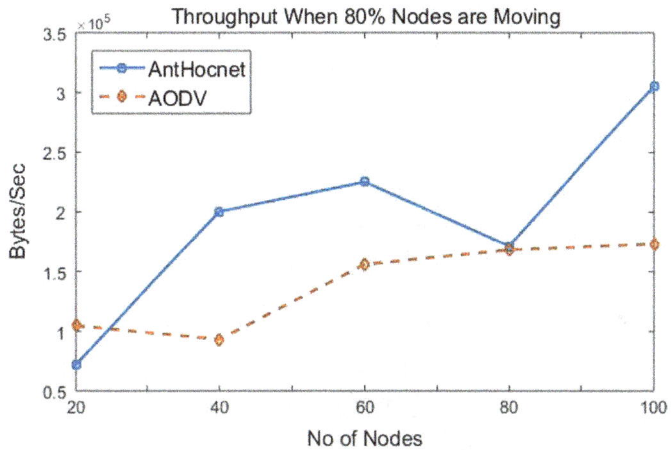

Fig. 9 Throughput when 80% nodes are moving

the node number is 80, the throughput of both routing protocols is more and less same.

From Figs. 6, 7, 8 and 9, it is seen that throughput of AntHocNet is better than AODV in maximum scenarios.

3.3 Average End-to-End (avg. e2e) Delay

Avg. e2e delay measured the average time taken for packets to reach destination from source node in a network. Figures 10, 11, 12 and 13 describe the comparison of avg. e2e delay when 20, 40, 60 and 80% nodes are moving, respectively.

When 20% nodes are moving, the avg. e2e delay of AntHocNet is less than AODV routing protocol. But when node number is 60 and 100, the avg. e2e delay of both routing protocols is near about same.

When 40% nodes are moving, the avg. e2e delay of AntHocNet is minimum here, but at the node number of 20 and 40 the avg. e2e delay is more or less same.

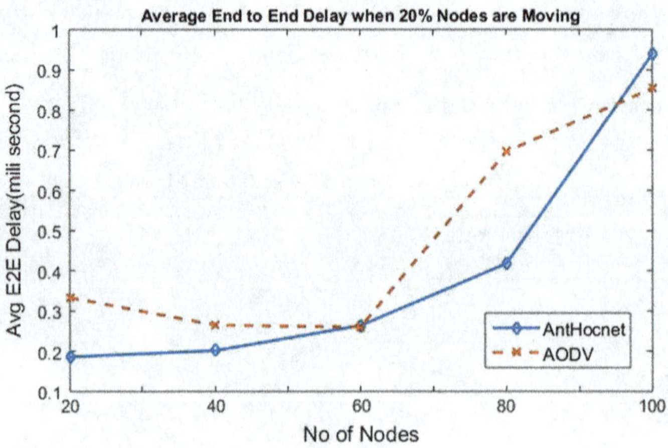

Fig. 10 Avg. e2e delay when 20% nodes are moving

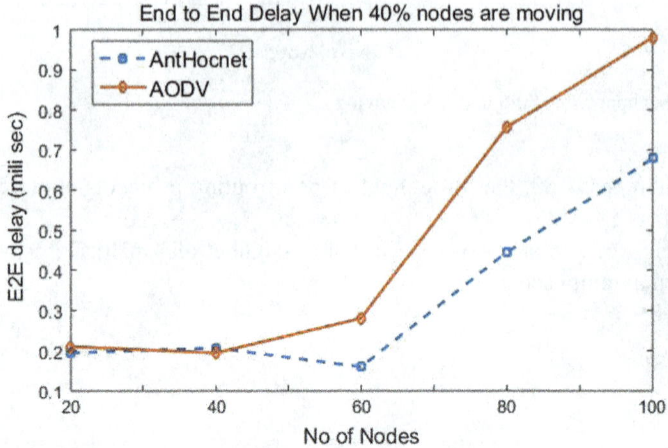

Fig. 11 Avg. e2e delay when 40% nodes are moving

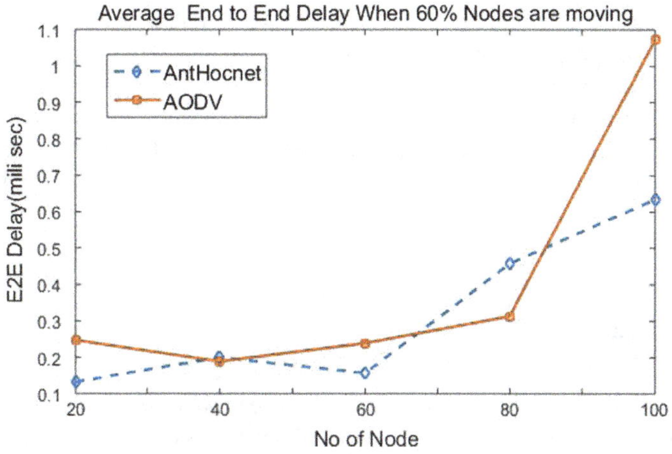

Fig. 12 Avg. e2e delay when 60% nodes are moving

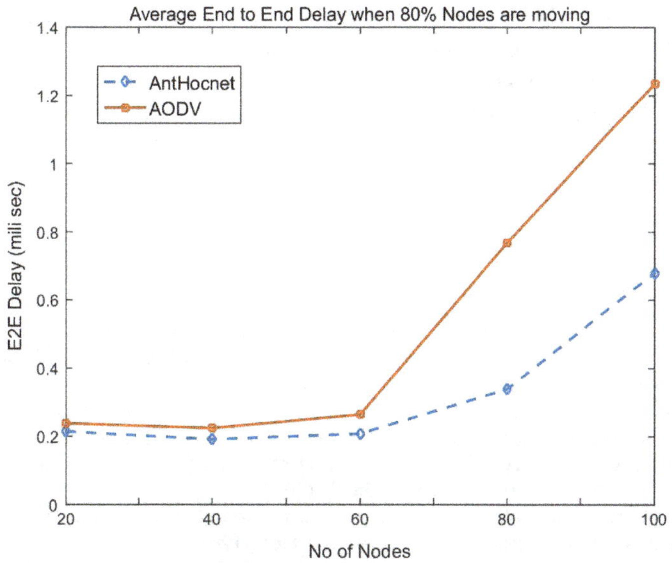

Fig. 13 Avg e2e delay when 80% nodes are moving

When 60% nodes are moving, the avg. e2e delay of AntHocNet and AODV is shown in Fig. 12. When the node number is 60, the average e2e delay of AntHocNet is little higher than AODV routing protocol, but rest of the cases avg. e2e delay of AntHocNet is less than AODV routing protocol.

When 80% nodes are moving, the avg. e2e delay is always less than AODV in every scenario.

From Figs. 10, 11, 12 and 13, it is seen that avg. e2e delay of AntHocNet is less than the avg. e2e delay of AODV routing protocol in maximum scenarios.

From Figs. 2, 3 , 4, 5, 6, 7, 8, 9, 10, 11, 12 and 13, it is seen that the performance of AntHocNet is always better than the convention routing algorithm (AODV). As AntHocNet routing protocol used metaheuristic approach to find the route to reach the destination so the performance of AntHocNet is better than AODV routing protocol here.

4 Conclusion

From the comparison of the performance, it is seen that the AntHocNet performs better than convention routing protocol (AODV) in terms of avg. e2e delay, throughput and total number of generated packets. In the few cases, AntHocNet behaves abnormally. The cause of abnormal behavior may be for other parameters like speed of node and mobility model, congestion of the route, etc. As the nodes are moving with a high speed (100 m/s), mobility model arranges the node position randomly. Overall, the performance of AntHocNet is satisfactory level than the conventional routing protocol like AODV.

For feature work, the AntHocNet routing protocol may be optimized for improving the performances of other various parameters in various scenarios.

References

1. Chakravarthy, C.K., Vinod Kumar, Y.: An integrated approach for quality of service (QoS) with security in VANETs. In: International Conference on Computer Networks and Communication Technologies. Springer, Singapore (2019)
2. Gunantara, N., Putra, N., Nyoman, I.D.: The characteristics of metaheuristic method in selection of path pairs on multicriteria ad hoc networks. J. Comput. Netw. Commun. (2019)
3. Solomon, A.A., Olusesi, A.T.: A model for self-adaptive routing optimization in mobile ad-hoc network. Int. J. Swarm Intell. Res. **10**(1), 58–74 (2019)
4. Mirjalili, S.: Ant colony optimisation. In: Evolutionary Algorithms and Neural Networks. Studies in Computational Intelligence, vol. 780. Springer, Cham (2019)
5. Kaur, S., Aseri, T.C., Rani, S.: QoS-aware routing in vehicular ad hoc networks using ant colony optimization and bee colony optimization. In: Proceedings of 2nd International Conference on Communication, Computing and Networking. Springer, Singapore (2019)
6. Yarinezhad, R., Sarabi, A.: A new routing algorithm for vehicular ad-hoc networks based on glowworm swarm optimization algorithm. J. AI Data Min. **7**(1), 69–76, Winter and Spring, Article 7 (2019). https://doi.org/10.22044/jadm.2018.6516.1765
7. Li, Y.Q., Wang, Z., Wang, Q.W., Fan, Q.G., Chen, B.S.: Reliable ant colony routing algorithm for dual-channel mobile ad hoc networks. Wirel. Commun. Mobile Comput. Article ID 4746020 (2018). https://doi.org/10.1155/2018/4746020

Design and Simulation of AND-OR-INVERT Logic for Photonic Integrated Circuits

Mahesh V. Sonth, Sanjaykumar Gowre, Nagshettappa Biradar and Basavaraj Gadgay

Abstract Optical fiber communication and photonics are the fastest developing field in the information processing and computational technology. The optical devices designed so far are not full optical processing systems rather they depend on the optical-electrical-optical (OEO) conversion switches in their system design. The necessity of these OEO conversion switches can be eliminated by designing all optical processing logic devices in photonic crystals (PhCs). The contrast ratio of the designed cascaded AOI logic gate is 11.15 dB achieved using Full-WAVE FDTD techniques.

Keywords Photonic crystals · OEO · FDTD · PWE · AOI logic

1 Introduction

Optics and photonics have received great attention by fundamental and applied researchers by the end of nineteenth century, the optical technology counterpart electronic semiconductor technology. The major challenging task in the field of communication system is to increase the processing speed, reduction in chip size, and low-power device design. Photonics is the research platform for the design and implementation of all optical devices for optical network systems. The physical science deal with light is refereed as photonics and PhCs are the Greek word *phos* means light and *krustallos* mean periodic arrangements [1, 2]. PhCs are the materials which have a periodicity dielectric constant in some particular dimensions and they are classified based on variation in refractive index as one-, two-, and three-dimensional PhCs [3].

M. V. Sonth (✉) · S. Gowre · N. Biradar
CMR Technical Campus, Hyderabad 501401, India
e-mail: maheshsonth@gmail.com

B. Gadgay
VTU PG Centre Kalaburagi, Gulbarga 585102, India

J. K. Mandal et al. (eds.), *Information, Photonics and Communication*, Lecture Notes in Networks and Systems 79,
https://doi.org/10.1007/978-981-32-9453-0_6

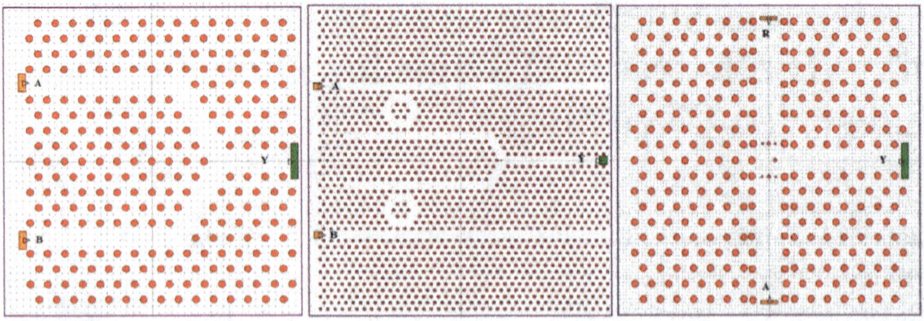

Fig. 1 PhC lattice structure of **a** AND gate, **b** OR gate, **c** NOT gate

All optical logic gates can be accomplished with 2-D PhCs through the use of PhCs waveguides and cavities. The various defects can be created in 2-D PhCs waveguide to guide electromagnetic radiation with small losses and high efficiency.

2 Structural Design Constraints

2.1 *AND Gate*

The all optical AND gate [4, 5] is designed in 2-D PhCs with hexagonal lattice structure of width 16 and length 19 silicon rods in air background medium shown in Fig. 1a. The lattice constant $a = 0.54$ µm, radius of silica rod $r = 0.108$ µm, and free space wavelength of 1.55 µm is set for the time domain analysis. The Gaussian stimulus power input is applied at the input of the structure A and B and the output is recorded at monitor placed at the output terminal Y.

2.2 *OR Gate*

The all optical OR gate [6] is designed in 2-D PhCs with hexagonal lattice structure of width 40 and length 49 silicon rods in air background medium shown in Fig. 1b. The lattice constant $a = 0.54$ µm, radius of silica rod $r = 0.108$ µm, and free space wavelength of 1.55 µm is set for the time domain analysis. The Gaussian stimulus power input is applied at the input of the structure A and B and the output is recorded at monitor placed at the output terminal Y.

2.3 NOT Gate

The all optical NOT gate [4, 7] is constructed in 2-D PhCs using hexagonal lattice structure as shown in Fig. 1c. The square lattice structure all optical NOT gate with input as A, R, and Y is output. 2-D PhCs with utility array layout width and length of 31 × 31 silica rods is in air background medium. The lattice constant $a = 0.54$ μm, radius of silica rod $r = 0.108$ μm, and free space wavelength of 1.55 μm is set for the time domain analysis.

2.4 AND-OR-INVERT Logic Gate

The PhC-based all optical AND-OR-INVERT (AOI) logic gate is constructed in 2-D PhC using the concept of cascaded logic structure shown in Fig. 2. One of the most tedious tasks is to obtain the high transmission efficiency in case of cascade logic. In this paper, we have presented cascade logic of AOI function using 2-D PhCs. The crystal lattice structure with a utility array width and length 40 × 39 silica rods is in air background medium. The lattice constant $a = 0.54$ μm, radius of silica rod $r = 0.108$ μm, and free space wavelength of 1.55 μm is set for the time domain analysis.

Photonic Bandgap: It is obtained by plane wave expansion (PWE) method results with transverse electric and transverse magnetic mode. The TE mode of PBG is ranging from 0.32 to 0.48 a/λ and TM mode of PBG is ranging from 0.86 to 0.956 a/λ for free space wavelength of 1550 nm. The contrast ratio r_c is computed by the logarithmic ratio of optical transmitted high power to the low power.

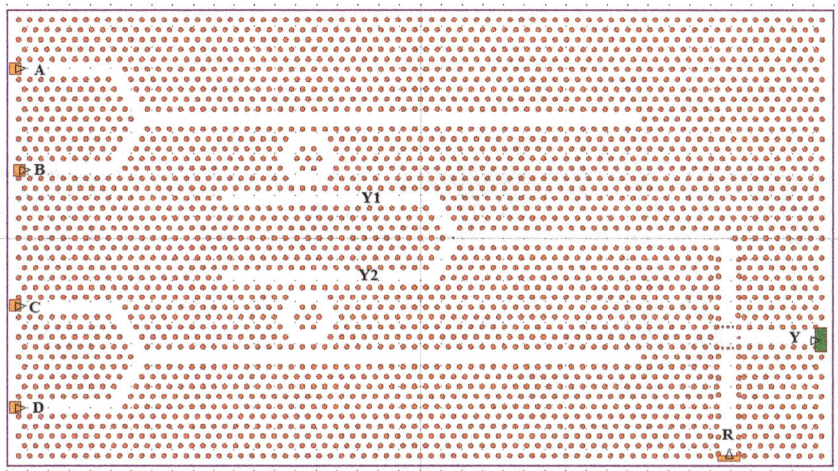

Fig. 2 PhC lattice structure of AND-OR-INVERT logic

$$\text{Contrast Ratio} = 10\log\left(\frac{P_1}{P_0}\right) \tag{1}$$

3 Simulation Results and Discussion

3.1 *AND Gate*

The all optical AND gate with input stimulus A and B is applied with power of 1 a.u. The output monitor records with high power output of 0.75 (arbitrary unit) a.u. and in remaining cases, it is observed with low power reading of about 0.18 a.u. Figure 3 shows the Full-WAVE FDTD simulation results of all optical AND gate in 2-D PhCs. Table 1 indicates the truth table of logical AND gate with practical output power and contrast ratio.

3.2 *OR Gate*

The all optical OR gate with input stimulus A and B is applied with power of 1 a.u. The output monitor records with high power output of 0.89 a.u.; if any one input is applied with a power of 1 a.u., then the output scope recorded with high power of 0.70 a.u., and if no input is applied means, then we will observe zero output in the monitor. Figure 4 shows the Full-WAVE FDTD simulation results of all optical OR gate in 2-D PhCs. Table 2 indicates the truth table of logical OR gate with practical output power.

3.3 *NOT Gate*

The input A with 1 a.u. power and reference input R are applied with 1 a.u. power, the low power (0.062 a.u.) coupled to output waveguide. Similarly, when we applied with input a low power and reference high power, due to coupling, high power (0.65 a.u.) will be coupled to output waveguide shown in Fig. 5. The extinction ratio of about 10.21 dB is achieved. Table 3 indicates the truth table of logical NOT gate with practical output power and contrast ratio.

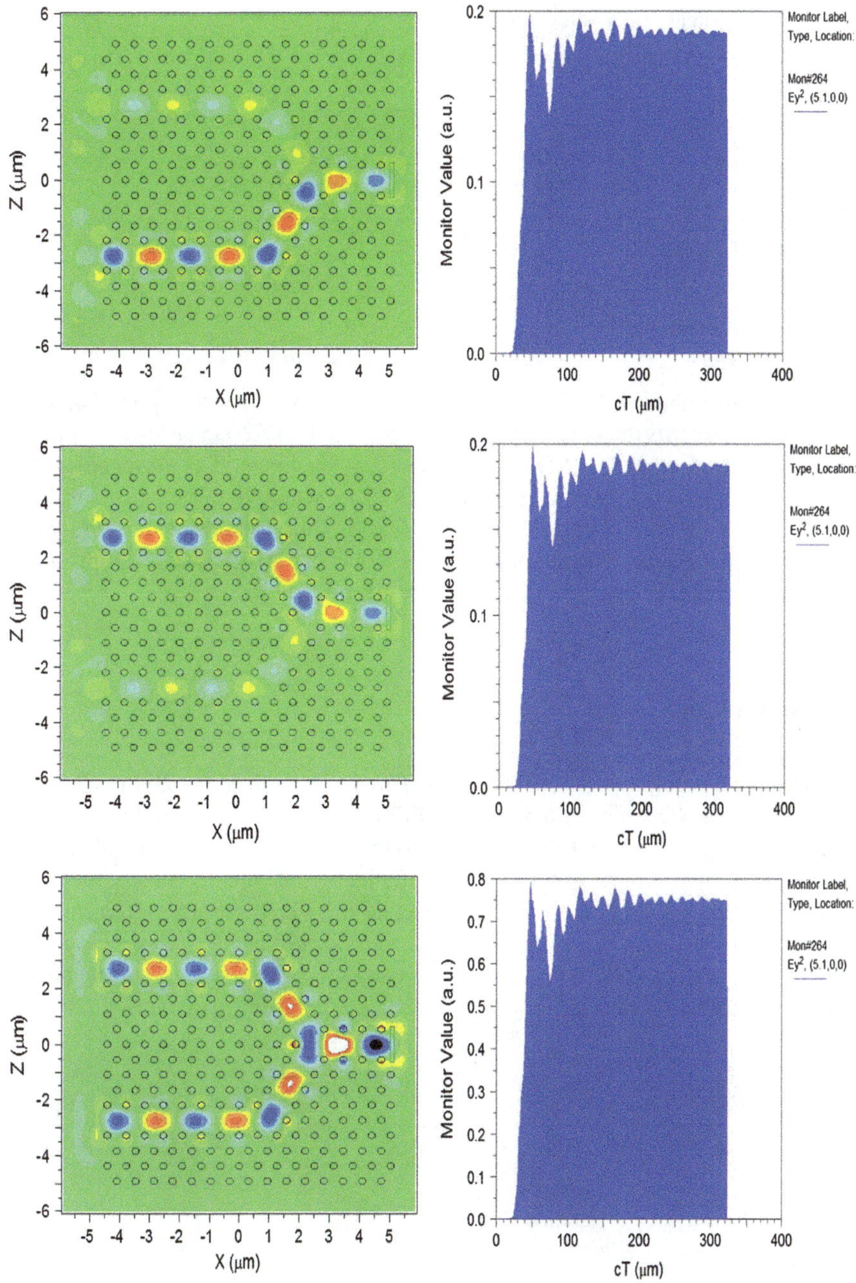

Fig. 3 Full-WAVE simulation of 2-D PhC AND gate

Table 1 FDTD simulation results of AND gate

Input		Theoretical output	Practical output (a.u.)	Contrast ratio (dB)
A	B	Y	Y	r_c
0	0	0	0	6.19
0	1	0	0.18	
1	0	0	0.18	
1	1	1	0.75	

3.4 AOI Logic Gate

AOI gate with inputs A, B, C, and D are given with an appropriate power of 1 a.u. and output Y records the output. The AOI logic with two AND gates with inputs A, B and C, D are designed in 2-D PhCs produces two outputs on its waveguides. The output of AND gates is used as a input of OR gate which intern cascaded to NOT gate generates the output of AOI gate. The high output produces when any one of AND gate input is zero. When both the inputs of any one AND gates are high, output goes to low. Table 4 shows the simulation results of AOI gate.

4 Conclusion

The basic gates such as AND, OR, and NOT logic are built in the 2-D PhCs lattice structure. The AND lattice structure in 2-D PhCs yielding a transmission efficiency of 75% logic high output and 18% of logic low output with a contrast ratio of 6.19 dB. The OR gate with maximum transmission efficiency for logic high output reported with 89%. The NOT gate designed in 2-D PhCs reported with 65% transmission efficiency for logic high output and 6% of transmission efficiency for logic low output with a contrast ratio of 10.21 dB. Finally, the cascaded AOI logic function maintained an average transmission efficiency of 72% with a contrast ratio of 11.15 dB.

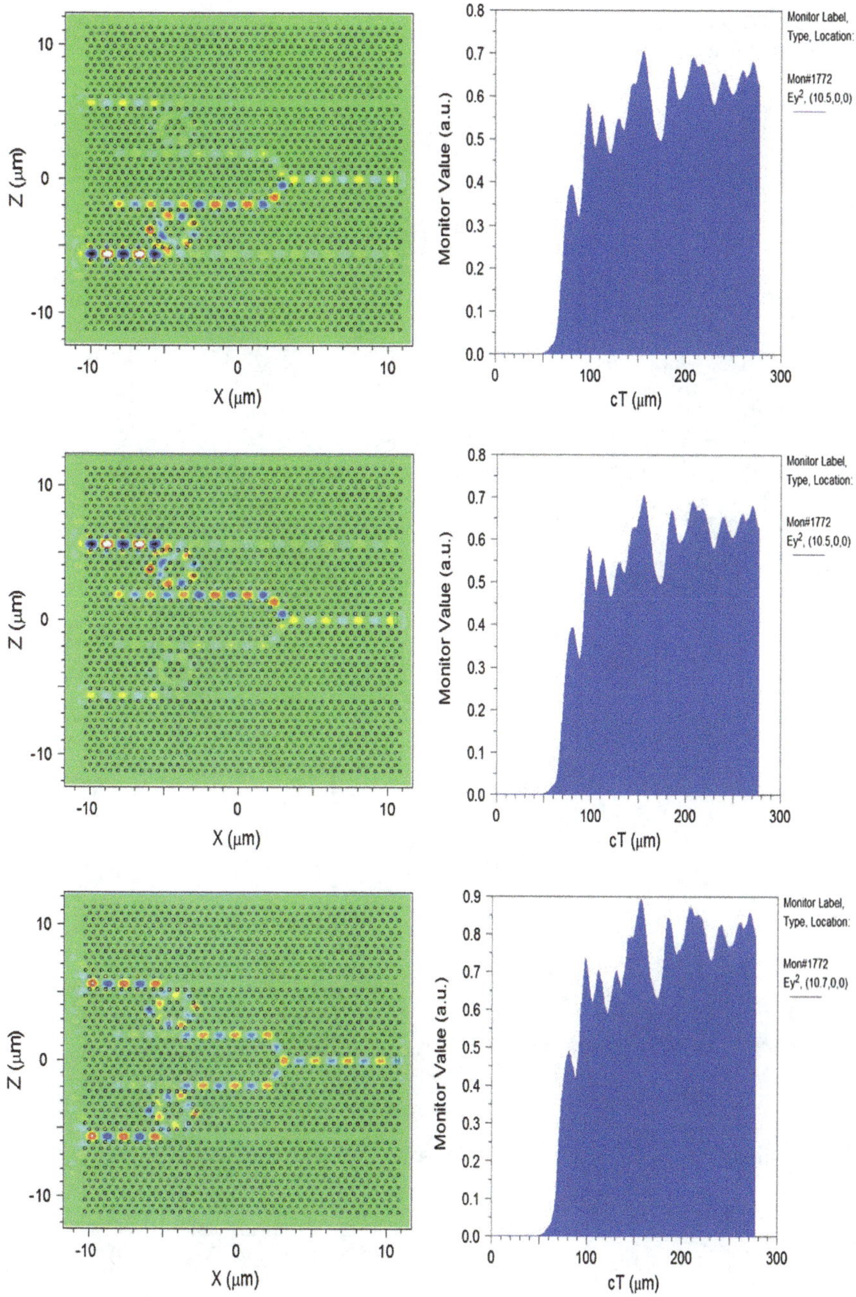

Fig. 4 Full-WAVE simulation of 2-D PhC OR gate

Table 2 FDTD simulation results of OR gate

Input		Theoretical output	Practical output (a.u.)
A	B	Y	Y
0	0	0	0
0	1	1	0.70
1	0	1	0.70
1	1	1	0.89

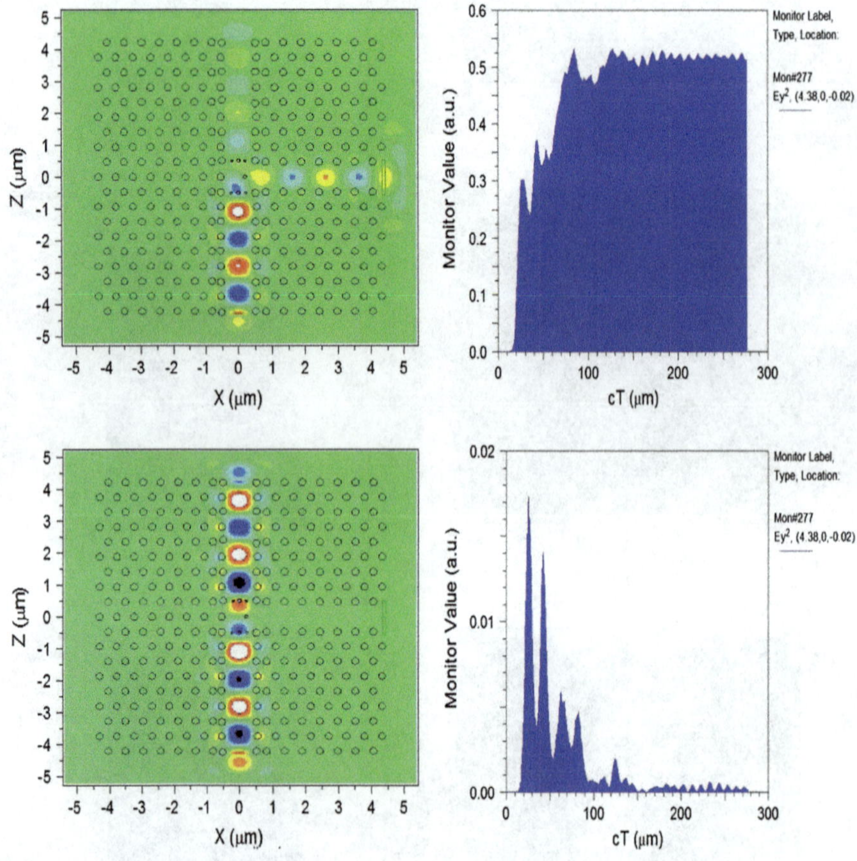

Fig. 5 Full-WAVE simulation of 2-D PhC NOT gate

Table 3 FDTD simulation results of NOT gate

Input	Theoretical output	Practical output (a.u.)	Contrast ratio (dB)
A	Y	Y	r_c
0	1	0.650	10.21
1	0	0.062	

Table 4 FDTD simulation results of AOI logic gate

Input				Logic output	Practical output (a.u.)	Contrast ratio (dB)
A	B	C	D	Y	Y	r_c
0	0	0	0	1	0.81	11.15
1	0	0	0	1	0.64	
0	1	0	0	1	0.72	
0	0	1	0	1	0.64	
0	0	0	1	1	0.72	
1	1	0	0	0	0.07	
0	0	1	1	0	0.07	
1	1	1	1	0	0.13	

References

1. Yablonovitch, E.: Inhibited spontaneous emission in solid-state physics and electronics. Phys. Rev. Lett. **58**(20), 2059–2062 (1987)
2. John, S.: Strong localization of photons in certain disordered dielectric superlattices. Phys. Rev. Lett. **58**(23), 2486–2489 (1987)
3. Pennec, Y., Vasseur, J.O., Djafari-Rouhani, B., Dobrzyński, L., Deymier, P.A.: Two-dimensional phononic crystals: examples and applications. Surf. Sci. Rep. **65**(8), 229–291 (2010)
4. Gupta, M.M., Medhekar, S.: All-optical NOT and AND gates using counter propagating beams in nonlinear Mach–Zehnder interferometer made of photonic crystal waveguides. Optik Int. J. Light Electron Opt. **127**(3), 1221–1228 (2016)
5. Macker, A., Shukla, A.K., Dubey, V.P.: A novel design of all optical AND gate based on 2-D photonic crystal. In: International Conference on Emerging Trends in Computing and Communication Technologies (ICETCCT), pp. 1–3. IEEE (2017)
6. Rahmani, A., Asghari, M.: An ultra-compact and high speed all optical OR/NOR gate based on nonlinear PhCRR. Optik Int. J. Light Electron Opt. **138**, 314–319 (2017)
7. Shaik, E.H., Rangaswamy, N.: Realization of XNOR logic function with all-optical high contrast XOR and NOT gates. Opto-Electron. Rev. **26**(1), 63–72 (2018)

Identifying Influential Nodes Based on Network Topology: A Comparative Study

Anindita Raychaudhuri, Subhasis Mallick, Ankit Sircar and Shalini Singh

Abstract Recently, the attention has been increased to the study of the connectivity properties and to the topology of the complex networks. This paper studies the relationship between the influential node and the topological structure of a network. Identification of influential nodes receives paramount interest as it is important for many real-world applications to identify strategically important nodes in different networks including social networks. Several node centrality measures are there for the identification of influential node. To overcome the limitations of well-known centrality measures like degree centrality, closeness centrality, and betweenness centrality, two more techniques are considered; one based on local properties of nodes (local-area centrality) and another based on global properties of the network (structural centrality). This paper investigates the role of local properties and position of the node with respect to the entire network for influential node identification with these algorithms. The experimental result shows that local-area centrality and structural centrality algorithms are able to identify the influential node with less computation time and effectiveness compared to other algorithms.

Keywords Network analysis · Node centrality · Influential node

A. Raychaudhuri (✉) · S. Mallick · A. Sircar · S. Singh
B.P. Poddar Institute of Management and Technology, 137 V.I.P Road, Kolkata 700052, India
e-mail: aninrc@gmail.com

S. Mallick
e-mail: subhasis.mallick@gmail.com

A. Sircar
e-mail: ankit.sircar@gmail.com

S. Singh
e-mail: shalinis597@gmail.com

J. K. Mandal et al. (eds.), *Information, Photonics and Communication*,
Lecture Notes in Networks and Systems 79,
https://doi.org/10.1007/978-981-32-9453-0_7

1 Introduction

In recent days, identifying the influential node in a network is very significant since it plays an important role in the areas of the computer network as well as social networks such as viral marketing, online recommendation, healthcare communities, disease spreading, expert finding, and rumour spreading [1–9]. It plays a significant role to control the outbreak of epidemics, to carry out successful advertisements and so on. The influence of a node in a network can be measured by the node centrality values. Any node is considered more central in a network if it is connected in a network such a way that it can circulate any message quickly to the greater section of the network. Through the analysis of various real-world networks, the researchers will be able to explore the connectivity properties of complex network systems to find many interesting structural properties of the network. Identifying node centrality has both theoretical and practical significance for many real-world applications including online social network. The degree centrality is a local and simple metric for identifying node centrality but of little relevance [3–6]. Some familiar global metrics such as closeness centrality and betweenness centrality are there for better results. But because of very high-computational complexity, they are not suitable to apply in very large-scale networks [3–6]. Therefore, the design of an impressive ranking method for identifying node centrality is still an open problem. Though degree centrality performs well for small network, local centrality is more suitable for the large-scale network. Degree centrality has little relevance as it is only has local information. A node may have many neighbours, but it may reside close to the periphery of a network. So a node with neighbours having a higher sum of degrees is more central or influential. So if degrees of next neighbours are cumulatively added, it adds more non-local information to calculate centrality. In [3], the authors have proposed a semi-local centrality measure to identify influential nodes. In this paper [3], they have considered neighbour and next neighbours. In our proposed approach, this definition has been extended and k-next nearest neighbours have been considered for better result termed as local-area centrality. We also propose a global structural information-based centrality measure which is based on the concept of k-shell decomposition [1] termed as structural centrality. The analysis shows that local-area centrality and structural centrality approach are identifying influential nodes with similar efficiency but with less computation time.

2 Background

In recent times, a vast change has been observed in the field of research on networks [10]. Initially, limited interests have been observed mainly from mathematical society, but recently it has received a huge amount of attention from many branches of sciences. In recent years, different methods have been put forward for discovering node strength in complex networks. The node centrality information helps to build applications such as controlling message spreading in social networks, promotion of a new product through viral marketing, ranking reputation of authors, etc. The control over influential nodes may stop rumour propagation in social networks, save from communication failure, and reduce the spread of epidemic. Specifically in biological networks, vital nodes are often promising targets for controlling the disease outbreaks.

A node is central if it has more connections, it can reach all the neighbours more speedily, and it controls the flow between the others. Three different measures have been formalized by Freeman based on these features which are degree, closeness, and betweenness [11]. Numerous network topology-based ranking schemes have been suggested for identifying influential nodes in real networks [7].

2.1 Degree Centrality

For a network $G = (V, E)$ with $n = |V|$ nodes and $m = |E|$ edges, degree centrality of node v is defined as the number of v's neighbours [4]. The graph G can be presented as an adjacency matrix $A = \{a_{uv}\} \in \{0, 1\}^{n \times n}$. If two nodes u and v are adjacent then $a_{uv} = 1$, otherwise $a_{uv} = 0$.

$$C_{\mathrm{D}}(v) = \sum_{u=1}^{n} a_{vu}$$

2.2 Betweenness Centrality

For a network $G = (V, E)$ with $n = |V|$ nodes and $m = |E|$ edges, the betweenness centrality of node v, denoted by $C_{\mathrm{B}}(v)$ is [4]

$$C_{\mathrm{B}}(v) = \sum_{s \neq v \neq t \in V} \frac{\sigma_{st}(v)}{\sigma_{st}}$$

Here σ_{st} is the number of shortest paths between nodes s and t and $\sigma_{st}(v)$ denotes the number of shortest paths between s and t which pass through node v.

2.3 Closeness Centrality

Closeness centrality was defined as the inverse sum of the shortest distances to all other nodes from a focal node. Closeness of node v is defined as [4]

$$C_c(v) = \frac{1}{\sum\limits_{t \in V/v} d_G(v,t)}$$

where $d_G(v,t)$ is the geodesic distance between v and t.

2.4 Local-Area Centrality

A node is more central in its locality if it has more neighbour. But if we consider next nearest neighbour of the node, then the centrality value changes and more non-local (global) information contribute in centrality measure.

The local-area centrality of node v, denoted by $C_L(v)$ is defined as

$$C_L(v) = |N_k(v)|$$

where $N_i(v)$ represents the set of neighbours of v with distance less or equal to i.

So $N_1(v)$ is the set of immediate neighbours of v

$$N_k(v) = N_1(v) \bigcup_{i_1 \in N_1(v)} N_1(i_1) \bigcup_{i_2 \in N_2(v)} N_1(i_2) \ldots \bigcup_{i_k \in N_k(v)} N_1(i_k)$$

Algorithm 1: Identification of the most influential node using local-area centrality approach

1: Start

2: Identify the nodes and links in the network.

3: Build the graph

4: Calculate the degree of each node

5: For each node initialize a list NODE_COVERED with the node

6: For i = 1 to k do

6_1: for each node do

6_1_1: Identify the neighbors of the current node of NODE_COVERED

6_1_2: Add these neighbors in NODE_COVERED

7: The node with maximum number of nodes in NODE_COVERED is most influential

8: Stop.

Local-area centrality can be defined as the sum of immediate neighbours and k-next level of neighbours [4]. Algorithm 1 shows the steps to calculate local-area centrality of a network.

2.5 Structural Centrality

A node which can be accessed from more nodes quickly is strategically more important than a node with more nearest neighbours. A node is structurally central in a network if it can send any message to other nodes quickly, or in other words, it is connected to the majority of the nodes in the shortest path.

To measure structural centrality, the concept of *k*-shell decomposition [1] is used. The algorithm starts by removing the lowest degree nodes first and add in a list. Then after removing associated edges of the removed node, the graph is reconfigured. This process continues until the graph is left with a node. That node will be central to the graph. The nodes in the list are ordered in ascending order of the structural centrality. This algorithm is described in Algorithm 2.

Algorithm 2: Identification of the most influential node using structural centrality approach

1: Start
2: Identify the nodes and links in the network.
3: Build the graph.
4: Calculate the degree of each node
5: Check the minimum degree of the graph
6: for all the nodes with the minimum degree of the graph.
6_1: Remove the node and add in a list REMOVED_LIST
6_2: Remove corresponding edges
6_3: Reconfigure the graph
7: If any node left in the graph Go to step 5.
8: Reverse the REMOVED_LIST to find the most influential node in top
9: Stop.

An example network with 17 nodes is given in Fig. 1. It shows the degree of node 5 is higher, but node 12 is structurally more central. Betweenness centrality

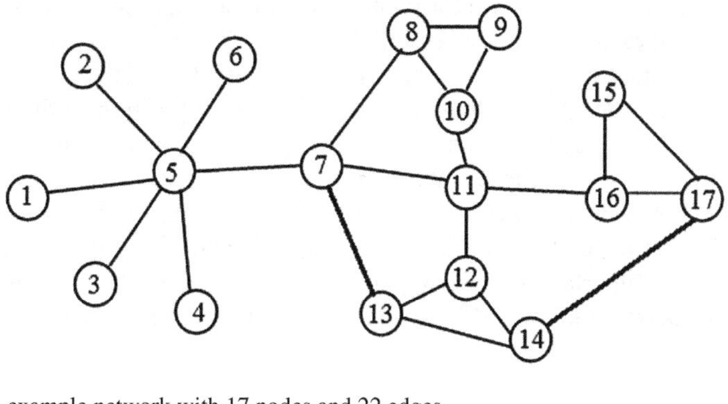

Fig. 1 An example network with 17 nodes and 22 edges

value is higher for node 7. Closeness centrality rank is highest for node 7. Therefore, identification of influential nodes not only depends on centrality algorithms, but also it is depended on the network topology.

3 Objective

In this work, we are primarily concerned about identifying central nodes in a network with feasible computational effort or less execution time. Nodes which are central in a network are strategically important in reality. In a network, the most central node will be identified as the most influential node of the network. Edges between nodes are assumed undirected and of the same weight that is all links are equally influential to spread information. Ranks of the nodes are identified based on their centrality value and less rank indicated more strength. Existing centrality measures are either not considering global structure of the network or taking high computation time which becomes infeasible for large data. To obtain the goal, the following objectives are considered:

- Frame the technique to measure local-area centrality based on semi-local information and perform empirical study based on large data sets.
- Frame the technique to measure structural centrality based on the global information of the network and perform an empirical study with a large data set.
- Compare the performance of degree centrality, closeness centrality, and betweenness centrality with local-area centrality and structural centrality technique with respect to properties of the network.
- Analyse how the rank varies with different techniques with node properties with large network data.

4 Analysis of the Properties of Centrality Measures

All the centrality measures have different limitation and application area based on the type of network and data size [1–9].

Degree of a node is the major property which helps to identify connection and association of a node. But it is not providing the overall position of the node with respect to the entire graph. So to deal with diverse requirements different centrality measures are proposed in the literature. Degree centrality deals with the degree of the node only, so it varies with the variation of the degree of the node. Local-area centrality is also depended on the degree. But along with degree the neighbours influence is also important to it. Betweenness and closeness centrality deals with the shortest path between all nodes, so it gives emphasize on entire network structure.

In betweenness centrality, no shortest paths are passing through many nodes. So their centrality value becomes 0 and hence difficult to identify their influential rank. Closeness centrality considers the shortest paths of all nodes to that node, so it cannot compute centrality for disjoint graphs. These limitations are overcome in structural centrality; it deletes the node with less degree first. So after deletion of the node which is structurally more central stays and receives higher centrality value. But as it deletes node iteratively, eliminated nodes do not have a contribution to all sections of the network. It is able to compute central nodes with less computational effort.

Centrality algorithms can be compared with respect to computational cost. Degree centrality requires calculating a total number of neighbours for all nodes, so it is $O(n^2)$ where n is the number of the node. Betweenness and closeness centrality deals with all possible shortest paths. Floyd's algorithm which finds all possible shortest paths has complexity $O(n^3)$ [3]. Local-area centrality requires degree of neighbours which requires $O(n^2)$ computation. If k-next levels of nodes are considered, then the computation multiplies by the scalar k. Structural centrality also deals with minimum degree nodes with computational complexity $O(n^2)$. It will iterate the maximum difference of degree times which is fixed for a network. Table 1 shows a comparative analysis of different centrality algorithms.

The graph in Fig. 2 shows the relation with degree and rank of different centrality algorithm. The graph considers first 50 influential nodes of the network constructed from *emailnet* data set. The degree centrality rank of a node varies with the degree of the node. Local-area centrality shows some dependence on the degree. Betweenness and closeness centrality ranks also show little dependency on degree, but structural centrality has almost no relation with the degree.

Table 1 Comparison of centrality algorithms

Centrality type	Degree of node	Computational complexity	Limitation
Degree centrality	Major impact	$O(n^2)$	Do not consider global position
Closeness centrality	Less impact	$O(n^3)$	Not applicable for disconnected network
Betweenness centrality	Less impact	$O(n^3)$	Same centrality value for many nodes
Local-area centrality	Less impact	$O(n^2)$	More emphasis on the local structure
Structural centrality	No impact	$O(n^2)$	Many nodes are eliminated

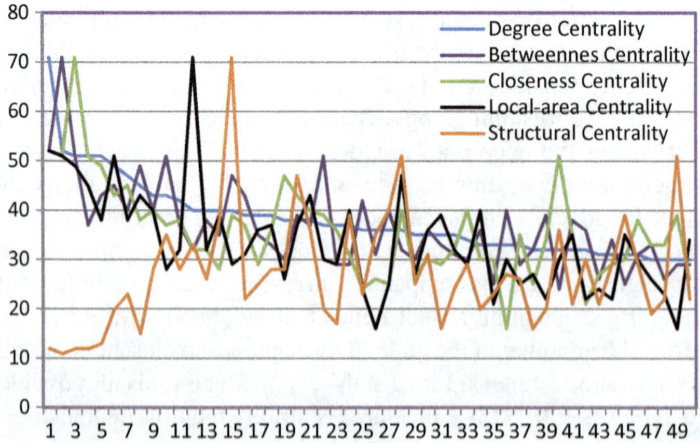

Fig. 2 Graph shows degree and rank co-relation for different centrality algorithms on emailnet data set; *x*-axis represents the rank of influential node and *y*-axis represents the degree

5 Empirical Study

5.1 Data

To evaluate the performance of the different centrality measures, we perform experiments using several real network data sets. Data sets are taken from different domains. All data sets are undirected. The data sets are collected from the Koblenz Network Collection [12]. We have used the data set of email communication network of University Rovira i Virgili in Spain as "Emailnet" data set, power grid data of the Western States of the USA as "Powergrid" data set, friendships between users of the website hamsterster.com as "Friendship" data set and collaboration network of authors of scientific papers from the arXiv's Astrophysics as "Co-author" data set. All the data sets contain at least two columns representing adjacent nodes, which are used to build the graph.

5.2 Experimental Result

Experiments are performed using NetworkX package of Python. To illustrate the performance of different centrality measures, we have implemented the algorithms based on different network data sets [12]. Outputs stored the list of nodes according to their increasing rank of centrality. To test their effectiveness, the node coverage is identified by different algorithms on different data sets. In the first iteration, the influential nodes can reach to their neighbours, and in the next iteration, they reach to the neighbour of neighbours. Table 2 shows the output of the first four iterations of

Table 2 Node coverage by most influential node in different centrality algorithms

Data set [12] (number of node)	Iteration	Node covered				
		Degree centrality	Closeness centrality	Betweenness centrality	Local-area centrality	Structural centrality
Co-author (18,771)	1	505	419	363	419	86
	2	5923	6065	5889	6065	1758
	3	14,598	14,659	14,640	14,659	10,528
	4	17,328	17,353	17,332	17,353	16,481
Friendship (1858)	1	273	273	273	149	70
	2	1148	1148	1148	1200	888
	3	1691	1691	1691	1720	1625
	4	1760	1760	1760	1768	1753

most central node of co-author and friendship data. Tables 3 and 4 shows the output of the first four iterations of the two most central and three most central nodes, respectively.

Experimental result shows that central nodes in structural centrality start with nodes with low degrees, but coverage is comparable with closeness centrality and betweenness centrality with less computational time. The powergrid network has no prominent central nodes, so it is giving poor result in all algorithms. Table 4 shows that the performance of structural centrality algorithm is very close to high-cost closeness centrality algorithm with respect to node coverage.

To compare the performance, we have identified the common nodes within the first 100 most central nodes of different algorithms on different data set [12] and tested their intersections.

Table 3 Node coverage by two most influential nodes in different centrality algorithms

Data set [12] (number of node)	Iteration	Node covered				
		Degree centrality	Closeness centrality	Betweenness centrality	Local-area centrality	Structural centrality
Co-author (18,771)	1	899	678	698	764	106
	2	7831	7113	7730	8266	1877
	3	15,410	15,087	15,396	15,625	10,676
	4	17,483	17,395	17,332	17,514	16,506
Friendship (1858)	1	300	373	373	373	160
	2	1193	1367	1367	1367	1051
	3	1692	1733	1733	1733	1675
	4	1760	1769	1769	1769	1756

Table 4 Node coverage by three most influential nodes in different centrality algorithms

Data set (number of node)	Iteration	Node covered				
		Degree centrality	Closeness centrality	Betweenness centrality	Local-area centrality	Structural centrality
Co-author (18,771)	1	1101	1010	937	1010	106
	2	8403	8873	8842	8875	1877
	3	15,639	15,790	15,833	15,790	10,676
	4	17,516	17,525	17,522	17,525	16,506
Friendship (1858)	1	410	390	500	411	334
	2	1288	1373	1489	1401	1232
	3	1694	1733	1756	1735	1695
	4	1763	1769	1772	1769	1760

Table 5 shows that the performance of local-area centrality is close to closeness centrality. Among the first 100 most influential nodes, these two algorithms have maximum common nodes. Influential nodes in local-area centrality also have a large intersection with degree centrality. The intersection of structural centrality with other centrality varies widely with the size of the network.

6 Conclusion and Future Scope

In this paper, we have implemented two centrality algorithms one based on local information and one based on the global position of the nodes. The performances of these two algorithms are compared with respect to three existing benchmark centrality algorithms. We have executed with moderate size data and found a significant difference in execution time.

However, some limitations are there which must be further expanded. In emailnet data, it is observed that the influential nodes are adjacent nodes. As a result, even if three most influential nodes have been used for spreading, they will show almost the same spreading capability as the most influential node has. This issue will be addressed in future work.

Table 5 Number of common nodes in 100 most influential nodes using different centrality algorithms

100 most influential nodes in	Data set [12] (number of node)	Number of common influential nodes in			
		Degree centrality	Betweenness centrality	Local-area centrality	Structural centrality
Closeness centrality	Co-author (18,771)	77	49	96	17
	Emailnet (1133)	70	71	93	49
	Powergrid (4941)	7	48	18	2
	Friendship (1858)	79	62	88	67
		Degree centrality	Closeness centrality	Local-area centrality	Structural centrality
Betweenness centrality	Co-author (18,771)	54	49	50	12
	Emailnet (1133)	75	71	66	39
	Powergrid (4941)	18	48	30	1
	Friendship (1858)	71	62	65	47
		Degree centrality	Closeness centrality	Betweenness centrality	Structural centrality
Local-area centrality	Co-author (18,771)	76	96	50	15
	Emailnet (1133)	64	93	66	47
	Powergrid (4941)	39	18	30	24
	Friendship (1858)	74	88	65	59
		Degree centrality	Closeness centrality	Betweenness centrality	Local-area centrality
Structural centrality	Co-author (18,771)	19	17	12	15
	Emailnet (1133)	56	49	39	47
	Powergrid (4941)	33	2	1	24
	Friendship (1858)	70	67	47	59

(continued)

Table 5 (continued)

100 most influential nodes in	Data set [12] (number of node)	Number of common influential nodes in			
		Degree centrality	Betweenness centrality	Local-area centrality	Structural centrality
		Closeness centrality	Betweenness centrality	Local-area centrality	Structural centrality
Degree centrality	Co-author (18,771)	77	54	76	19
	Emailnet (1133)	70	75	64	56
	Powergrid (4941)	7	18	39	33
	Friendship (1858)	79	71	74	70

References

1. Kitsak, M., Gallos, L.K., Havlin, S., Liljeros, F., Muchnik, L., Stanley, H.E., Makse, H.A.: Identification of influential spreaders in complex networks. Nat. Phys. **6**(11) (2010)
2. Zhu, J., Zhu, J., Ghosh, S., Wu, W. and Yuan, J.: Social influence maximization in hypergraph in social networks. In: IEEE Transactions on Network Science and Engineering (2018)
3. Chen, D., Lü, L., Shang, M.S., Zhang, Y.C., Zhou, T.: Identifying influential nodes in complex networks. Physica A **391**(4), 1777–1787 (2012)
4. Ma, Q., Ma, J.: Identifying and ranking influential spreaders in complex networks with consideration of spreading probability. Physica A **465**, 312–330 (2017)
5. Gao, S., Ma, J., Chen, Z., Wang, G., Xing, C.: Ranking the spreading ability of nodes in complex networks based on local structure. Physica A **403**, 130–147 (2014)
6. Liu, Y., Tang, M., Zhou, T., Do, Y.: Identify influential spreaders in complex networks, the role of neighborhood. Physica A **452**, 289–298 (2016)
7. Chintakunta, H., Gentimis, A.: Influence of topology in information flow in social networks. In: Proceedings of the 2016 Asilomar Conference on Signals, Systems and Computers, pp. 67–71, Pacific Grove, CA, USA, 6–9 Nov 2016
8. Lü, L., Zhang, Y.C., Yeung, C.H., Zhou, T.: Leaders in social networks, the delicious case. PLoS ONE **6**(6), e21202 (2011)
9. Liu, S., Jiang, C., Lin, Z., Ding, Y., Duan, R., Xu, Z.: Identifying effective influencers based on trust for electronic word-of-mouth marketing: a domain-aware approach. Inf. Sci. **306**, 34–52 (2015)
10. Newman, M.E.J.: Networks: An Introduction. Oxford University Press, Oxford (2010)
11. Das, K., Samanta, S., Pal, M.: Study on centrality measures in social networks: a survey. Soc. Netw. Anal. Min. **8**(1) (2018)
12. KONECT data collection. http://konect.uni-koblenz.de/networks

Towards a Novel Cross-media Encryption-Cum-Obfuscation Technique

Dipnarayan Das and Sumit Gupta

Abstract The safeguarding of sensitive information has always been the mission of researchers and cryptographers since time immemorial. People all around the world have always felt the dire need to establish and ensure information security, i.e., to ascertain that crucial and confidential information is kept secure from any unwanted access and/or attack. Moreover, it is also necessary to provide access to confidential data to legitimate and authentic users after confirming login credentials. Cryptography is a well-known technique for maintaining confidentiality, integrity, authentication, and non-repudiation of messages, whereby the sender encrypts the plain text to generate ciphertext, sends it across a communication medium which on the receipt is decrypted by the authentic receiver. This paper proposes a novel cross-media encryption-cum-obfuscation technique where the encryption technique will obfuscate itself by changing its media. To simulate the proposed obfuscation technique in the real world, a specific hardware setup is requisite at both the terminals. So, the transmitter and receiver hardware circuitry is smartly designed to dynamically change the signatures and achieve enhanced security. Thus, it can be said that the proposed technique is robust against the traditional cryptanalysis methods and can ensure information security by obfuscating the data.

Keywords Cryptography · Encryption · Transceiver · Obfuscate · Masking

1 Introduction

Cryptography, a Greek word which means 'secret writing', refers to the science and art of transforming messages so as to make them secure and immune to attacks. It is generally used for entity authentication, i.e., to authenticate the sender and

D. Das · S. Gupta (✉)
Department of Computer Science and Engineering, University Institute of Technology, The University of Burdwan, Golapbag (North), Burdwan 713104, West Bengal, India
e-mail: sumitsayshi@gmail.com

D. Das
e-mail: dipnarayan.das35@gmail.com

J. K. Mandal et al. (eds.), *Information, Photonics and Communication*, Lecture Notes in Networks and Systems 79,
https://doi.org/10.1007/978-981-32-9453-0_8

receiver of the message to each other. It involves the transformation of the original message (called as plaintext) into a ciphertext using an encryption algorithm, whereas a decryption algorithm transforms the ciphertext back into plaintext. The sender uses an encryption algorithm, and the receiver uses a decryption algorithm [1]. There are different types of encryption techniques available in the literature. The encryption process uses an algorithm to scramble or encrypt data and then uses a key for the receiving party to unscramble or decrypt the information. Cryptanalysis is the study of ciphers and cryptosystems to find weaknesses in them that would allow access to the information without knowing the key or algorithm. Frequency analysis is a technique that is generally used to crack a cipher. Generally, people who try to decrypt a message have to study about the frequency of letters (or groups of letters) in a cipher text. Because some letters occur more often than others, the frequency of letters can reveal parts of the encrypted message. While this method was effective in cracking old encryption methods, it is ineffective against modern encryption techniques.

In this paper, the proposed system comprises different sections for implementing the encryption technique and for developing a transceiver system. The proposed system is based upon the client-to-client communication architecture. Both the clients, viz. sender (transmitter) and receiver, have the transceiver system for ensuring bidirectional communication. In the input terminal, two things can be done either an animation can be inputted which will be transferred through a sound wave for hiding the real identity of the client or an audio (or sound) can be inserted in real-time by its SD storage or system integrated mic.

The rest of the paper is organized as follows: Sect. 2 discusses the previous related works. Section 3 presents the proposed methodology in terms of system components required and the working principle of the system that uses the cross-media encryption technique to ensure information security. Section 4 discusses about the implementation part and the results so obtained. Section 5 concludes the paper by listing out the future scope of improvement that can be taken up.

2 Previous Related Works

Many papers on cryptography have been studied where different authors have proposed varied encryption techniques. Few of the most popular encryption techniques are discussed in this section.

Whole disk encryption refers to the encryption of an entire physical or logical disk or its full contents. It is generally applied on software, but tremendous work is undergoing to explore the benefits of the hardware-based disk encryption [2].

Encryption can be done either on a single-user file or a multi-user file. A file can be encrypted individually using a password or other key to create an encrypted virtual drive that ensures the protection of data against disclosure. Simultaneous access of encrypted information by multiple users is more complicated than that of a single user since it necessitates the use of either multiple keys or a shared key or password and has to deal with multi-user file locking issues.

While dealing with a database, encryption can be done at the application level which can encrypt the original information before storing it into the database. This is basically done using the built-in encryption functions. An encryption application can act as an intermediary between the application and the database and can encrypt or decrypt information at the time of writing or reading [3]. Some applications that are not designed for encryption do have few basic encryption functions [4].

There are two ways to encrypt an email firstly by encrypting an attached file using any single-file encryption process and secondly by encrypting an entire message. It requires the receiver to know the decryption process beforehand in order to get back the original file. Two commonly used email message encryption technologies are S/MIME and PGP that allow for digital signing of email without encrypting it. The digital signing process allows the receiver to be certain that a message was unaltered while passing through the medium, but it does not protect against eavesdropping.

Secure Sockets Layer/Transport Layer Security (SSL/TLS) is commonly used to encrypt Web traffic in communication channel or transit. It also authenticates the communication between clients and servers done via email. SSL/TLS can also be used for tunneling so as to encrypt other forms of network transmission that do not have their own encryption features.

IPSec is a more general form of network traffic encryption that operates at a more basic layer than SSL or SSH. As IPSec requires common configuration between the two communicating computers, so it is generally used within an organization but not across the Internet [5].

Wireless security deals with only encrypting information between the computer and the wireless access point. Thus, the only protection from snooping on a wireless network is ensured. Wired Equivalent Privacy (WEP) and WiFi Protected Access (WPA) are two of its well-known variants. WEP is no longer considered a secured protocol. WPA is more robust but due to its shortcomings, WPA2, an updated version has been released [6].

Secure voice or secure speech or ciphony is a term in cryptography meant for the encryption of voice communication over a range of communication types such as radio, telephone, or IP [7, 8].

None of the previous methods [9–13] are capable of converting its media to obfuscate. When any attacker will not know what to attack, then it will be very tough to decipher the target system.

3 Proposed Methodology

3.1 System Components

The proposed system (see Fig. 1) contains different embedded equipment which are described as follows:

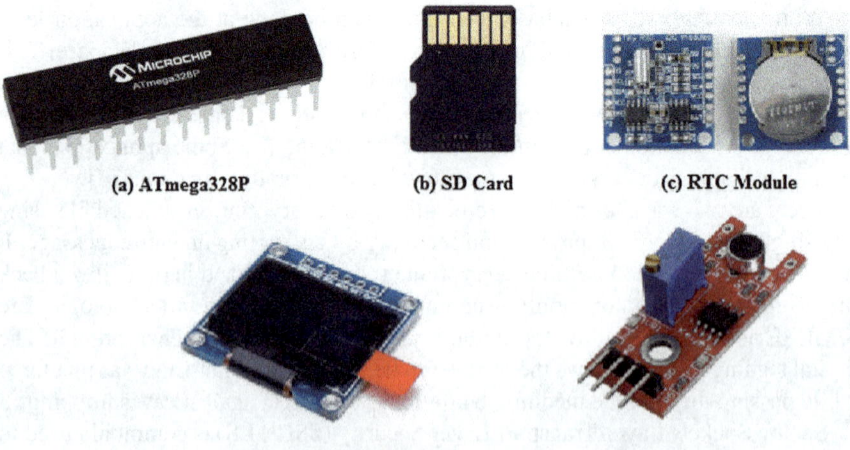

(a) ATmega328P (b) SD Card (c) RTC Module

(d) Organic LED Display (e) Condenser Mic

Fig. 1 Components of the proposed system

1. **ATmega328P:** This is a 28 pin microcontroller which is used to process the audio signal to visualize and vice versa.
2. **SD Card:** The Secure Digital (SD) card is used to store the audio files or the animations needed to be transmitted.
3. **RTC Module:** The real-time clock (RTC) module is used to synchronize the audio signal for sampling and correct formation of signals.
4. **OLED Display:** Organic light-emitting diode (OLED) display is used to see the image or for visualization.
5. **Condenser Mic:** The condenser mic is used to receive the audio signal transmitted through the air. Further, the wired input is also available.

3.2 Working Principle

To develop the proposed system, major emphasis has been given on domains such as encryption in cross-media, image, and video processing and speech and audio signal processing.

The proposed system is based on peer-to-peer communication architectural model which works as follows: Firstly, from the transmitter section, an image or a video is inputted which will be further converted into an analog signal depending upon the visual frame channels, bits per pixel or any audio which has already been encoded. After converting into audio signal, if it is sent to the visualizer, then it will again format the audio in the same way so as to visualize. In this case, the transmission media is sound, but the output is an image or video. In the case of video transmission, the audio of the video will also be encrypted. So now to mask the main raw signal from being hijacked, there are different masking techniques that can be used. The following discusses few popular masking techniques:

1. **Translation:** In this case (see Fig. 2), the audio has shifted its position to a different place to make it out of bound in the visual representation. The shift of positions will create ambiguity in the mind of the intruder or hacker.
2. **Scaling:** In the case of scaling (see Fig. 3), depending upon the display resolution, the intensity values are changed which will lead the result into an incorrect visual representation. Thus, until and unless the hacker knows which display resolution is used to visualize it, s/he will not be able to perform correct visualization.
3. **Stereo Input:** In case of stereo input (see Fig. 4), the transmission takes place in the wired medium. Here, different channels are present which will hide the target audio signal channel.

In the receiver section, the audio signal will be sampled to reform the visualization. To make it more obfuscated, the transceiver systems are capable for commanding each other to change the receiving frequency of the audio signal. So at any random moment, the transmitter can change its frequency to stop or prevent watching of outsiders. There are specific commands that can be set through different mediums, e.g., Google Mail and Outlook to make it more dynamic. In case of sending any command, pulse-width modulation is used for making the starting and ending quotation. For enhanced privacy, we can also use the ultrasound transmitter and receiver. Figures 5 and 6 show schematics of the transmitter circuit along with its two phases and Figs. 7 and 8 show schematics of the receiver circuit along with its phase.

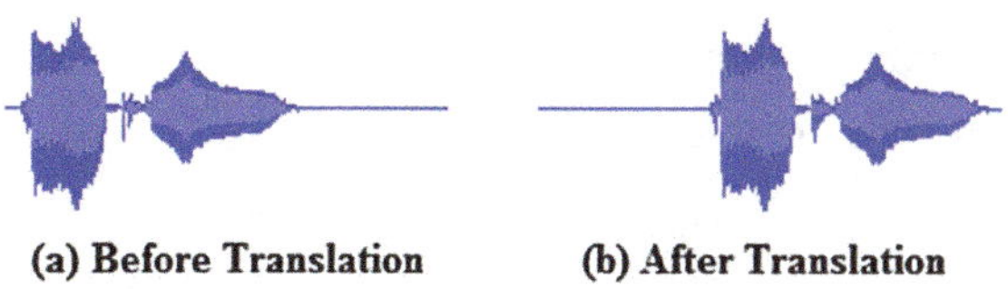

Fig. 2 Effect of translation

Fig. 3 Effect of 40% scaling on the signal

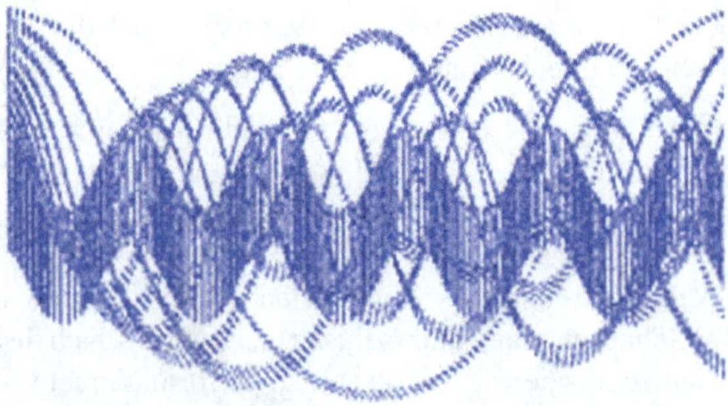

Fig. 4 Target signal is in stereo mode

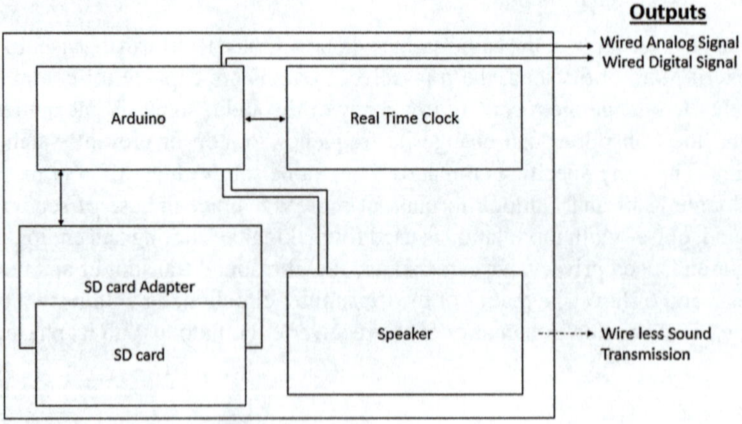

Fig. 5 Transmitter circuit of the proposed system

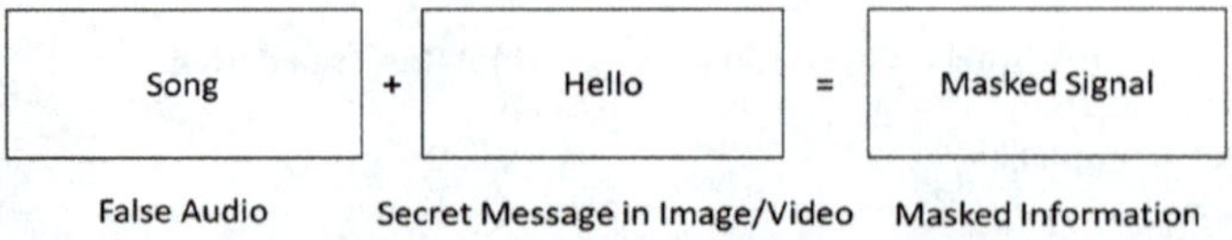

(a) Transmitter Phase 1.

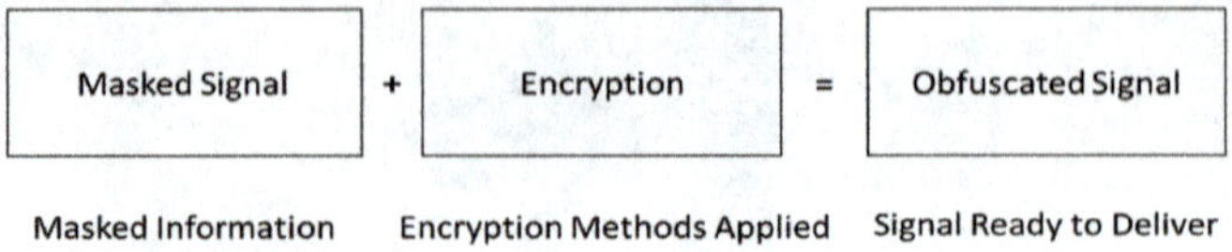

(b) Transmitter Phase 2.

Fig. 6 Phases of the transmitter

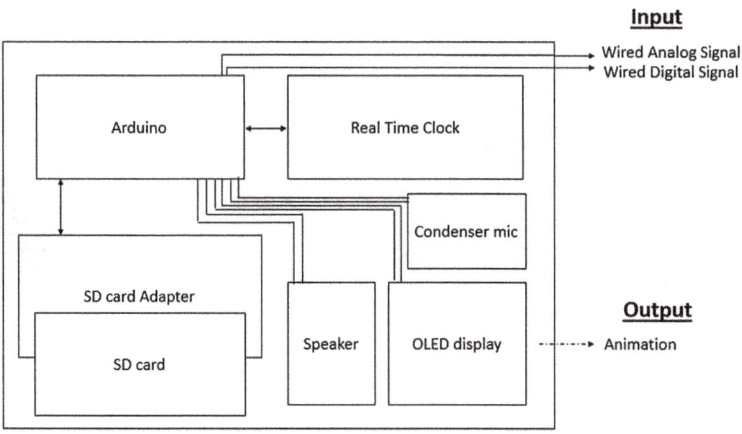

Fig. 7 Receiver circuit of the proposed system

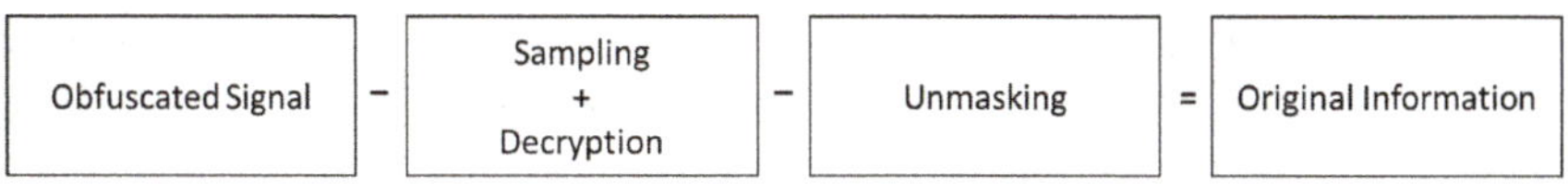

Fig. 8 Phase of the receiver

4 Implementation and Result

After sample testing, the system with stereo sound, it was observed that for one channel, the data is obfuscated by another text (see Fig. 9).

The original source file (either text or image or video) is sent through wired medium along with the randomly selected audio files (for instance, Song 1 and Song 2) which will obfuscate the original information to dupe the hacker. As the added audio signals have varying intensities, the hacker will tend to misunderstand the

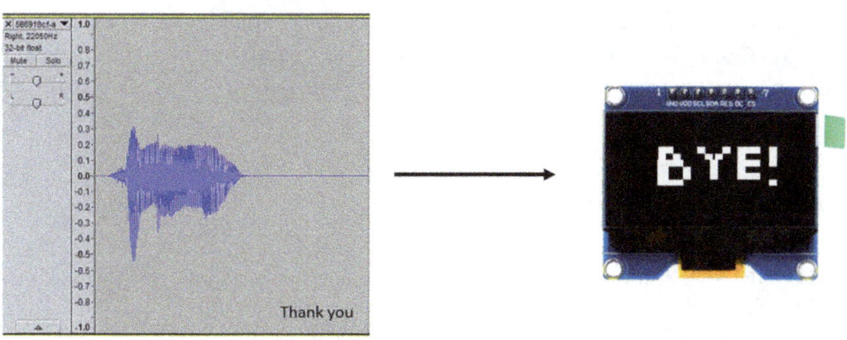

Fig. 9 Resultant image from a sound input

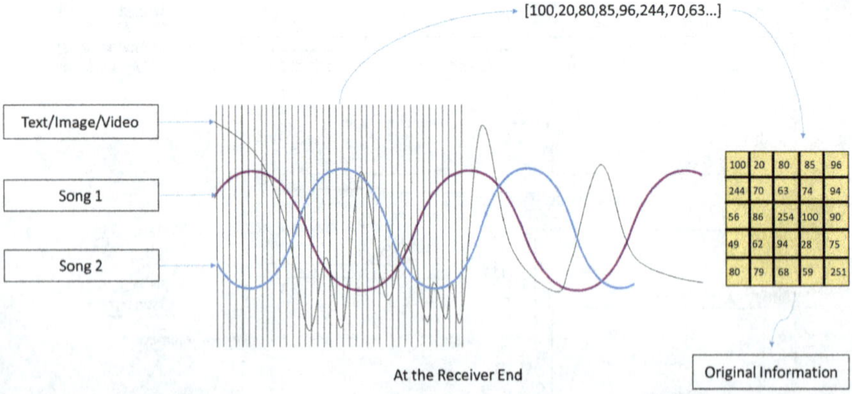

Fig. 10 Conversion process of an obfuscated signal into the original information

obfuscated signal as the original signal or message and thus the original data will get secured. Moreover, as the authentic receiver knows the actual stereo channel used for sending the original source information, s/he will sample it and get back the original information produced as 1D data and store these data into a 2D array. Thus, only the credible receiver will be able to receive the actual message and the hacker will misjudge the receiving signals and will not be able to interpret the information correctly. Figure 10 demonstrates the conversion process showing how an obfuscated signal (comprising the original message as text or image or video) gets converted into original information in a wired medium.

5 Conclusion and Future Work

Through this paper, a novel cross-media encryption-cum-obfuscation technique is proposed and the proposed system is shown to work in an automated fashion. The presented approach uses the concept of obfuscation to hide the data identity and thus proves to be better than most of the currently available exploit systems.

In the future, we aim to enhance the level of security by adding more levels of encryption to the proposed approach presented in this paper.

References

1. Forouzan, B.A.: Data Communications and Networking, 4th edn. McGraw-Hill Forouzan Networking Series, New York (2007)
2. Symantec Webpage. https://www.symantec.com/content/en/us/enterprise/white_papers/b-pgp_how_wholedisk_encryption_works_WP_21158817.en-us.pdf. Accessed 2018/12/10

3. Singh, P., Kaur, K.: Database security using encryption. In: International Conference on Futuristic Trends on Computational Analysis and Knowledge Management (ABLAZE) 2015, pp. 353–358. IEEE, Noida, India (2015)
4. Thales eSecurity Webpage. https://www.thalesesecurity.com/faq/encryption/what-application-encryption. Accessed 2018/12/10
5. Dhall, H., Dhall, D., Batra, S., Rani, P.: Implementation of IPSec protocol. In: Second International Conference on Advanced Computing & Communication Technologies, pp. 176–181. IEEE, Rohtak, Haryana, India (2012)
6. Lashkari, A.H., Danesh, M.M.S., Samadi, B.: A survey on wireless security protocols (WEP, WPA and WPA2/802.11i). In: 2nd IEEE International Conference on Computer Science and Information Technology, pp. 48–52. IEEE, Beijing, China (2009)
7. Albahrani, E.A.: A new audio encryption algorithm based on chaotic block cipher. In: Annual Conference on New Trends in Information & Communications Technology Applications (NTICT) 2017, pp. 22–27. IEEE, Baghdad, Iraq (2017)
8. Khalil, M.I.: Real-time encryption/decryption of audio signal. Int. J. Comput. Netw. Inf. Secur. (IJCNIS) **8**(2), 25–31 (2016)
9. Gandhi, R.A., Gonsai, A.M.: A study on current scenario of audio encryption. Int. J. Comput. Appl. **116**(7), 13–17 (2015)
10. Kaur, M., Kaur, S.: Survey of various encryption techniques for audio data. Int. J. Adv. Res. Comput. Sci. Softw. Eng. **4**(5), 1314–1317 (2014)
11. Gandhi, R.A., Gonsai, A.M.: Audio encryption with AES and blowfish. Int. J. Res. Appl. Sci. Eng. Technol. (IJRASET) **4**(11), 671–679 (2016)
12. Jawahir, A., Haviluddin, H.: An audio encryption using transposition method. Int. J. Adv. Intell. Inform. **1**(2), 98–106 (2015)
13. Tamimi, A.A., Abdalla, A.M.: An audio shuffle-encryption algorithm. Int. J. Adv. Intell. Inform. (IJAIN) **1**, 98–105 (2015)

Computing Electromagnetic Bandgap Structure of Metamaterial-Based 2D Photonic Crystal for TM Mode

Ratul Ghosh, Papri Chakraborty, Anwesha Adhikary and Arpan Deyasi

Abstract The electromagnetic bandgap of metamaterial-based two-dimensional photonic crystal structure with rectangular geometry is analytically computed by plane-wave expansion method. A wide range of DNG materials is considered for simulation purpose to calculate the width of bandgap (forbidden region) and their position coordinates in the band diagram. It has been established that complete bandgap formation is possible for TM mode, and the results are also compared with that obtained for positive refractive index materials with air holes of similar dimension. Peak point coordinates of 'equivalent' valence bands are notified for different DNG materials which will reveal the position shift of lowermost forbidden region. Existences of multiple such regions are confirmed for DNG materials which will speak for its candidature in utilizing as a multi-channel filter at a high-frequency range.

Keywords Metamaterial · Electromagnetic bandgap · TM mode · Photonic crystal · Two-dimensional structure · Forbidden region

1 Introduction

Photonic crystal is the building block of the twenty-first century where all-optical integrated circuit is considered as a possible replacement [1] of optoelectronic circuits

R. Ghosh · P. Chakraborty · A. Adhikary
Department of Electronic Science, A.P.C College, Kolkata, India
e-mail: ratulghosh777@gmail.com

P. Chakraborty
e-mail: papri.chakraborty1996@yahoo.co.in

A. Adhikary
e-mail: piu.anwesha@gmail.com

A. Deyasi (✉)
Department of Electronics and Communication Engineering, RCC Institute of Information Technology, Kolkata, India
e-mail: deyasi_arpan@yahoo.co.in

J. K. Mandal et al. (eds.), *Information, Photonics and Communication*,
Lecture Notes in Networks and Systems 79,
https://doi.org/10.1007/978-981-32-9453-0_9

owing to improved SNR, less area for floorplanning and reliability. One of the key components is the photonic crystal fiber [2], having significantly better characteristic than its present counterpart. Photonic crystal is already used in designing optical transmitter [3], receiver [4], sensor [5], switch [6], quantum information processing [7], waveguide [8], etc. The major advantage of making these various devices is the inherent nature of the structure of working as a filter due to the restriction of propagation of certain electromagnetic frequencies, while allow others to transmit [9]. This property is nomenclature as photonic bandgap [10], already utilized in making various antenna in different desired high-frequency domains. Metamaterials are nowadays employed in different antenna structures for enhanced SNR [11], and henceforth, investigation of their fundamental optical property is one of the primary concerns for theoretical researchers. Therefore, the computation of electromagnetic band structure is one of the most important parts of metamaterial-based photonic crystal from an application point of view.

Research on bandgap engineering in photonic crystal is started a few years ago when Karla published a few literatures [12, 13] on the orientation of dielectric voids and their shape. Lattice filling factors are also investigated recently [14, 15], and quality factors of those microcavities are computed as a function of the refractive indices of the constituent materials [16]. Systems are already preferred for applications in CWDM systems [17]. Effect of defects is investigated on this structure [18] and showed that these may improve the performance of the structure for some specific optical applications. Different numerical methods [19, 20] and algorithms [21] are tried to compute the near-accurate band structure, and very recently, single negative index (SNG) materials are proposed [22] to get multi-stopband characteristics. However, it is applied for 1D simple structure only, and maximum two stopbands are obtained for particular configurations. But at present, to the best knowledge of authors, researches are hardly carried out on DNG materials for 2D structure.

In the present paper, photonic bandgap is obtained for DNG material-based 2D structures where multiple stopband formations are shown over a wider range of refractive index variations. Corresponding peak points are noted and compared with positive refractive index materials. Plane-wave expansion method is used for computation purpose, and the simulation reveals the enormous possibility of making multi-channel stopband filter with DNG materials.

2 Mathematical Formulation

For any arbitrary medium, Maxwell's equation for wave propagation may be put into the following form:

$$\nabla \times \frac{1}{\varepsilon(r)} \nabla \times \vec{H}(r) = \frac{\omega^2}{C^2} \vec{H}(r) \tag{1}$$

The propagating mode in any periodic configuration may be expressed, according to Bloch's theorem, the sum of an infinite number of plane waves:

$$H(r) = \sum_{\vec{L}_i,\lambda} h_{L_i,\lambda} e^{i(\vec{k}+\vec{L}_i).\vec{r}} \hat{p}_\lambda \tag{2}$$

where λ is a positive integer, k denotes the wave vector, $\vec{L}$ is the reciprocal lattice vector in that space where the crustal structure is considered, $\hat{p}_\lambda$ is denotes two-unit axes normal to the direction of propagation $(\vec{k} + \vec{G})$, $h_{L,\lambda}$ is signifies the coefficient of the H component along $\hat{p}_\lambda$.

Applying Fourier transform, the dielectric function may be put into the form

$$\varepsilon(\vec{L}) = \frac{1}{V} \oiiint_{\Omega} \varepsilon(r) \exp(-i\vec{L}.\vec{r}) \tag{3}$$

where integration is performed over the entire unit cell, whose volume is defined by V. For computing the eigenvalue, we use Helmholtz's equation

$$\sum_{\vec{G}'} |k + L||k + L'|\varepsilon^{-1}(L - L')\begin{bmatrix} \hat{p}_2\hat{p}_2' & -\hat{p}_2\hat{p}_1' \\ -\hat{p}_1\hat{p}_2' & \hat{p}_1\hat{p}_1' \end{bmatrix}\begin{bmatrix} h_1' \\ h_2' \end{bmatrix} = \frac{\omega^2}{C^2}\begin{bmatrix} h_1 \\ h_2 \end{bmatrix} \tag{4}$$

where $\begin{bmatrix} \hat{p}_2\hat{p}_2' & -\hat{p}_2\hat{p}_1' \\ -\hat{p}_1\hat{p}_2' & \hat{p}_1\hat{p}_1' \end{bmatrix}$ matrix indicates the direction of the wave vector propagation at the time of lattice scattering inside first Brillouin zone, $\begin{bmatrix} h_1' \\ h_2' \end{bmatrix}$ is a matrix which satisfies for the all solutions possible.

Due to the symmetry of Eq. (4) w.r.t TE and TM modes of propagation, the total number of equations can be reduced from 2 N to N. Based on the simplification, we can rewrite eigen equation as

$$\sum_{G'} |k + L||k + L'|\varepsilon^{-1}(\vec{L} - \vec{L}') = \frac{\omega^2}{C^2} h(\vec{L}) \tag{5}$$

For the polarized incidence of e.m wave, both k and L are in the same plane, which leads to the following simplifications

$$\begin{bmatrix} \hat{p}_2\hat{p}_2' & -\hat{p}_2\hat{p}_1' \\ -\hat{p}_1\hat{p}_2' & \hat{p}_1\hat{p}_1' \end{bmatrix} = \begin{bmatrix} 1 & 0 \\ 0 & \cos(\phi - \phi') \end{bmatrix} \tag{6}$$

For TM mode, eigen equation is given by

$$\sum_{L'} |k+L||l+L'|\varepsilon^{-1}(\vec{L}-\vec{L}')h_1(\vec{L}') = \frac{\omega^2}{C^2}h_1(\vec{L}) \tag{7}$$

3 Results and Discussion

Using Eq. (7), we have computed and plotted bandgap structure for different DNG materials under TM mode of propagation. In Fig. 1, the band structure is plotted for different metamaterials, and corresponding photonic bandgaps are depicted. In Fig. 1a, profile is made for parallel nanorod (r.i $= -0.3$), whereas higher negative values are chosen for Fig. 1b and c. It is seen from the comparative analysis that width of the photonic bandgap varies w.r.t refractive index and thus corresponding midband frequency.

From the analysis, we have computed the width of the lowest two bandgaps and coordinates of equivalent valence bands. The results are represented in tabular form, and variation of bandgap widths are plotted with refractive index variations in Figs. 2 and 3, respectively. Table 1 shows the forbidden region data, and coordinates are given in Table 2. A few results for positive refractive index materials are also provided for comparison.

The following figures represent the variation of photonic bandgap widths and also the shift of peak points. It may be noted that for very low negative index, the difference between two passband increases which will help to prevent the aliasing error. This gap is generally low for positive index materials, i.e., the width of the guard band is comparatively lower. Also the equivalent valence band shifts toward higher wavevector axis with lowering of the index, whereas, for positive index materials, the shift is taking in an opposite direction. This speaks about the position of the first Brillouin zone. Also the length of the region is increased for lower negative index materials, whereas it is decreased for higher positive index materials. Figure 2 represents the change of lowermost bandgap widths, whereas Fig. 3 denotes the width of the Brillouin region. It may also be noted that Brillouin zone width remains constant for the very low negative index, which feature is totally absent in positive index materials.

4 Conclusion

Photonic bandgap width and peak point coordinates of the metamaterial-based photonic crystal are analytically evaluated using the plane-wave expansion method. The variations of the lowest two bandgaps are analyzed with negative refractive indices. The shift of coordinates indicates the position of bandgap, which is going toward higher frequency range as negative index decreases. Computation is performed for

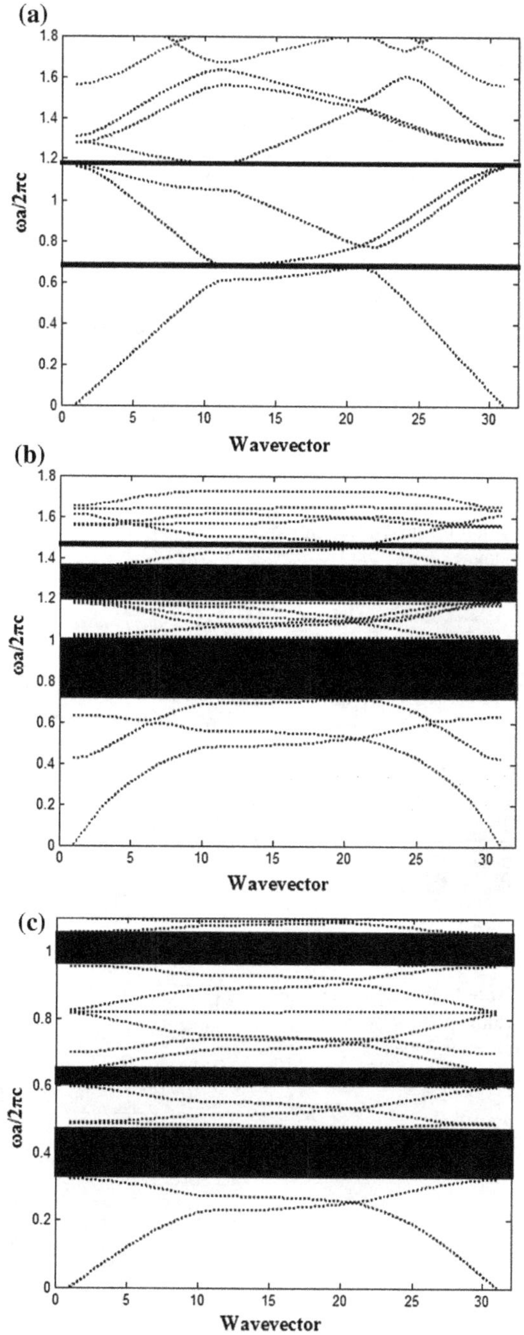

Fig. 1 **a** Photonic bandgap for DNG material with r.i = −0.3 for TM mode of propagation. **b** Photonic bandgap for DNG material with r.i = −7 for TM mode of propagation. **c** Photonic bandgap for DNG material with r.i = −25 for TM mode of propagation

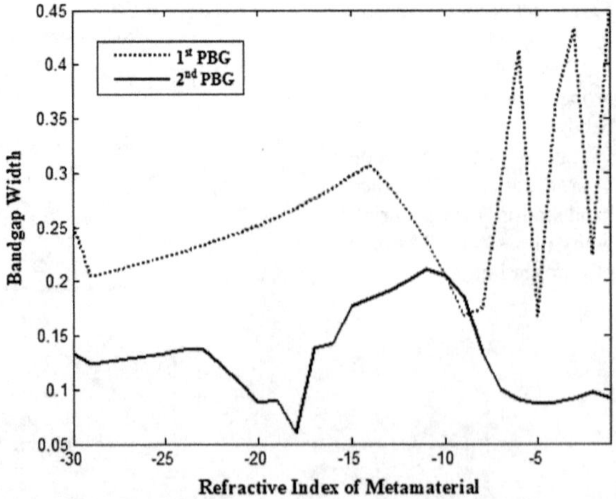

Fig. 2 Variations of the lowest two photonic bandgaps with negative index

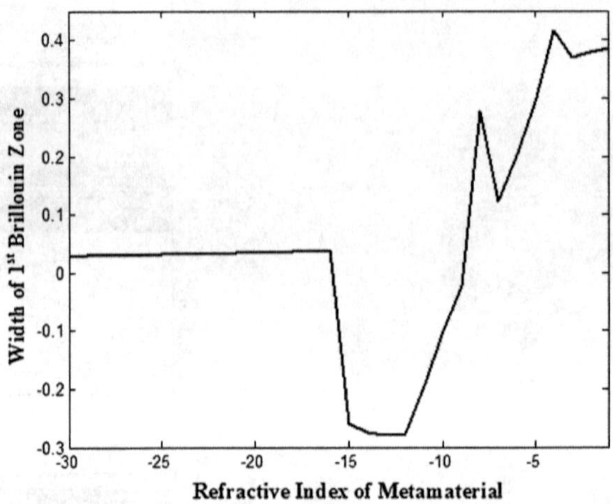

Fig. 3 Width of First Brillouin zone with negative index

Table 1 Photonic bandgap width variation with refractive index

Refractive index of material	Width of the first photonic bandgap	Width of the second photonic bandgap
−30	0.2509	0.1332
−25	0.2221	0.1341
−20	0.2523	0.0877
−15	0.2971	0.1764
−10	0.2048	0.205
−5	0.1675	0.202
5	0.3543	0.2152
10	0.282	0.1415
15	0.2318	0.1326

Table 2 Peak point coordinates of the equivalent valence band

Refractive index of the material	Coordinates of the first peak point	Coordinates of the second peak point
−30	(10, 0.202)	(21, 0.2309)
−25	(10, 0.223)	(21, 0.2544)
−20	(10, 0.2523)	(21, 0.287)
−15	(10, 0.297)	(21, 0.3362)
−10	(10, 0.3781)	(21, 0.4239)
−5	(10, 0.5978)	(21, 0.6508)
5	(10, 0.3852)	(21, 0.4636)
10	(10, 0.3014)	(21, 0.3559)
15	(10, 0.2551)	(21, 0.299)

rectangular lattice geometry. Quasi-bandgap formation is evident for the artificial materials as crossover point is observed for TM mode of propagation. The results are very important to design multi-channel bandpass filter using DNG materials.

References

1. Roelkens, G.: III-V-on-silicon photonic integrated circuits for optical communication and sensing. In: Frontiers in Optics: OSA Technical Digest, pp. FF5F.4 (2016)
2. Cavanna, A., Jiang, X., Taheri, M., Leuchs, G., Chekhova, M.V., Joly, N.Y.: Photonic crystal fiber designs for third-harmonic and photon triplet generation. In: Nonlinear Optics: OSA Technical Digest, pp. NTh2A.7 (2017)
3. Huang, S.C., Hong, K.B., Chiu, H.L., Lan, S.W., Lu, T.C.: Electrically driven GaAs-based photonic crystal lasers with ITO cladding layers. In: Proceedings of SPIE: High Contrast Metastructures VII, vol. **10542**, p. 105420R (2018)
4. Nozaki, K., Matsuo, S., Takeda, K., Sato, T., Fujii, T., Kuramochi, E., Notomi, M.: Photonic-crystal-based InGaAs photodetector connected to load resistor for receiver-less light-to-voltage conversion on chip. IEEE Photonics Conference (2014)
5. Babayiğit, C., Turduev, M., Giden, I.H., Bor, E., Kurt, H.: T-shape slotted photonic crystal based sensor with high sensitivity. In: 18th International Conference on Transparent Optical Networks (2016)
6. Dmitriev, V., Kawakatsu, M.N., Portela, G.: Compact optical switch based on 2D photonic crystal and magneto-optical cavity. Opt. Lett. **38**, 1016–1018 (2013)
7. Watcharakitchakorn, O., Silapunt, R.: Design and modelling of the photonic crystal waveguide structure for heat-assisted magnetic recording. Adv. Mater. Sci. Eng. **2018**, 8097841 (2018)
8. Adibi, A., Xu, Y., Lee, R.K., Loncar, M., Yariv, A., Scherer, A.: Role of distributed bragg reflection in photonic-crystal optical waveguides. Phys. Rev. B **64**, 041102 (2001)
9. Mao, D., Ouyang, Z., Wang, J.C.: A photonic-crystal polarizer integrated with the functions of narrow bandpass and narrow transmission angle filtering. Appl. Phys. B **90**, 127–131 (2008)
10. Das, B., Deyasi, A.: Effect of grating length on wave propagation inside paired DNG/Air Composition, CRC Press, Computational Science and Engineering, pp. 277–280 (2016)
11. Li, Z., Xue, Y.L., Shen, T.: Investigation of patch antenna based on photonic band-gap substrate with heterostructures. Math. Probl. Eng. **2012**, 151603 (2012)
12. Kalra, Y., Sinha, R.K.: Photonic band gap engineering in 2D photonic crystals. PRAMANA J. Phys. **67**(6), 1155–1164 (2006)

13. Kalra, Y., Sinha, R.K.: Modelling and design of complete photonic band gaps in two-dimensional photonic crystals. PRAMANA J. Phys. **70**(1), 153–161 (2008)
14. Mukherjee, S., Roy, A., Deyasi, A., Ghosal, S.: Computation of photonic bandgap in two-dimensional periodic triangular lattice for bandpass filter design. In: IEEE International Conference on Microelectronics, Computing and Communication, pp. 1–5 (2016)
15. Abdenacer, A., Fahima, A., Radouene, G., Fares, K.: The effect of radius and size on photonic bandgap of a triangular lattice of rods Si/Air-based 2D photonic crystals. Nanocon 2013 (2013)
16. Moukhtari, R., Hocini, A., Khedrouche, D.: Study of two-dimensional photonic crystal microcavities as a function of refractive index. Acta Phys. Pol., A **129**(4), 556–558 (2016)
17. Chhipa, M.K., Rewar, E.: Design and analysis of 2D hexagonal photonic crystal structure based channel drop filter for CWDM system. In: International Conference on Microwave, Optical and Communication Engineering (2015)
18. Ge, F.H., Zhang, J.H., Zhang, L.Q., Lu, L., Huang, X.K.: Effect of point and linear defects on bandgap properties in Triangular-Honeycomb structure photonic crystals. IOP Conference Series: Material Science Engineering **170**, 012005 (2017)
19. Najafi, A., Jalalkamali, M., Moghadamzadeh, S., Bolorizadeh, M.A.: Finite element method analysis of photonic crystal fiber band structure. In: Symposium on Photonics and Optoelectronics, pp. 1–4 (2010)
20. Salski, B.: The unfolding of bandgap diagrams of hexagonal photonic crystals computed with FDTD. Progr. Electromagnetics Res. M **27**, 27–39 (2012)
21. Hanif, A.G., Arima, T., Uno, T.: Finite-difference frequency-domain algorithm for band-diagram calculation of 2D photonic crystals composed of Debye-type dispersive materials. IEEE Antennas Wirel. Propag. Lett. **11**, 41–44 (2012)
22. Yeh, D.W., Wu, C.J.: Analysis of photonic band structure in a one-dimensional photonic crystal containing single negative materials. Opt. Express **17**(19), 16666–16680 (2009)

Inter-Vehicular Information System (IVIS) on IOT Platform

Soumyadip Chatterjee

Abstract Here I report the synthesis of inter-vehicular information system (IVIS) on IOT platform keeping in mind the factors like road ethics and road safety issues. Inter-vehicular information system is often called as inter-vehicular smart system. This project is made for the road unit team of the traffic department. IVIS can monitor 24*7 driver's cardiac condition, alcohol consumption, smoking inside the car, interior/exterior temperature of the car, physical damage made to the car, speed driving, and relative sea-level pressure around the car along with its altitude and carbon monoxide emission level of the car. It is not only a low-cost system but also an eco-friendly system too. Those features will be monitored on an IOT platform which will make my system wireless and portable both. This technology is mainly made to maintain the road safety ethics and side-by-side monitoring of driver's health and few physical parameters of the car. This proposed system will be very useful for the multi-national companies like OLA, UBER, or any ride-based companies. Up to a certain extent, it can be helpful for many company's delivery boys too.

Keywords Vehicular ad hoc network · Internet of things · BMP 180 (Bosch)

1 Introduction

The inter-vehicular information system (IVIS) is an automated communication network embedded electronically in the vehicle. This multifunction system displays its output to provide a graphical user interface for human–machine interaction, embedded sensors for the information amalgamation, a dispensed multiprocessor computer system for information processing, and a digital communications channel. Main battle tank M1A2 is an example of IVIS which offers many features along with benefits to the operators. In a moving situation, it can support essential graphic and digital data as per required to accomplish its mission; hence, it is fully automated. Previously, IVIS supports MANET (Mobile Ad hoc Network) information system to provide

S. Chatterjee (✉)
B.P. Poddar Institute of Management and Technology, 137, V.I.P. Road, Poddar Vihar, Kolkata 700052, India
e-mail: soumyadipchatterjee700@gmail.com

J. K. Mandal et al. (eds.), *Information, Photonics and Communication*, Lecture Notes in Networks and Systems 79,
https://doi.org/10.1007/978-981-32-9453-0_10

battle command capability and automated mobile combined arms force integrations. An Ad hoc network is an assembling of mobile nodes, which forms a short-term network without the assistance of central network/admin or any commercially based support network devices. As nodes have a short transmissions range, so each node searches for the aid from its nearby nodes in forwarding and receiving packets of data. Hence, the nodes can be treated as host and routers both. Walky Talky is an exemplar of MANET. By nature, these types of network are suitable for the situations where either no fixed groundwork exits or installing the network is not possible. Ad hoc network has various applications in military, emergency, and sensor network. The unique features of these protocols are their ability to track down routers in spite of a dynamic topology [1, 2].

Since the last decade, mobile communication has evolved so much that it made automobile industrial product to exchange valuable data/information between devices and substations. This is the new prototype for the exchange of information on a real-time basis. With all new advancements in the world of technology, there came an idea of vehicular ad hoc networks (VANET). While coming in contact over a wireless medium, VANET refers to the networks created for a special purpose for different moving vehicles where it can exchange their useful information among each other. At that very moment, a network is built between vehicles and other support devices acting as a node in the network. Now a node can be a vehicle or a substation or any other connecting devices. Information present in a node is relayed to all other nodes and after relaying the set of data received, the following transmission of data occurs. This transmission will continue until the data is fully received by the end users. It is an open network where there is no limitation for nodes to join and leave the network. This communication can be expanded between devices as much as required. These new ideas of VANET equipped with onboard sensors are implemented in new vehicles which are being launched in the market. VANET is a variation of the MANET system. Nodes in MANET have the networking capabilities and can communicate without the aid of central admin/network, whereas nodes in VANET have the freedom to set foot in and out of the network. This inter-vehicular communication has enhanced traffic mobility and lowered road accidents. It will also help to detect road conditions and emergency situations too [3, 4].

VANET can be characterized by the following parameters:

i. *Dynamic topology*—The physical parameters of a vehicle like its direction and its speed always changes constantly; therefore, a static topology is not efficient.
ii. *Sporadic connectivity*—Due to dynamic topology, the interconnection between devices changes frequently for exchanging information which can disconnect anytime.
iii. *Motility patterns*—Vehicles follow a certain pattern to move which is generally a function based on traffic signals, speed limits, etc.
iv. *Unlimited power and storage*—The nodes in VANET architecture are capable of processing a huge amount of power and storage capacity which makes nodes to be free in order to exchange their data with consumable power negligible or storage wastage.

v. *Onboard sensors*—VANET presumes that the nodes are not often provided with onboard sensors for transmitting information to other devices or nodes. VANET is an important part of intelligent transport systems as information being relayed among the vehicles or any other supportive devices [5].

Now, VANET can be classified into four sub-parts:

a. Vehicle-to-Vehicle (V2V) communications consist of a wireless network where automobiles convey messages to each other with information. This data can be speed, location, direction of travel, braking, and loss of stability.
b. Emerging wireless technologies for vehicle-to-station (V2S) communications such as dedicated short-range communication (DSRC) are promising to reduce the number of lethal roadway accidents by providing early warnings. Here the data's exchange is between vehicle and its nearest station (RSU-Road Side Unit).
c. Vehicle-to-infrastructure (V2I) is the wireless exchange of data between vehicles and road infrastructure. It is enabled by a system of hardware, software, and firmware. V2I is typically wireless and bidirectional. Infrastructure components such as road signs, lane markings, and traffic lights can wirelessly provide information to the vehicles and vice versa.
d. Vehicle-to-pedestrian communication approach encompasses a set of the large scale of road users including people walking, passengers stepping in, and out of the buses and trains and people riding bicycles, etc. Here it is a one-to-one communication between vehicles and its end users (pedestrians/passengers).

IVIS is advantageous to societal people are:

- Wireless and portable.
- Cheap and efficient.
- Eco-friendly.
- Battery-powered system.
- Its demand will rise simultaneously with 5G commercialization.

2 Proposed System

MQ-3 is the semiconductor sensor for alcohol gas, MQ-2 is the semiconductor sensor for combustible gas, and MQ-9 is the semiconductor sensor for carbon monoxide gas. BMP 180 is the digital barometric pressure sensor of Bosch Sensortec. Heartbeat module is an electronic device that is used to measure the heart rate. All these modules are connected with Arduino mega which is a microcontroller board based on the ATmega2560. BMP 180 has the ability to measure temperature, atmospheric pressure, and altitude simultaneously. MQ 2,3,9 have a direct application with their usage. Apart from direct application-based usage of BMP 180, I have used its pressure-sensing capability to detect speed driving and car collision both. Similarly, the heartbeat module is implemented to monitor driver's cardiac condition (Fig. 1).

Initially for the testing purposes, I have written the code in Arduino and got the output in the serial plotter. Then the whole code was uploaded in NodeMcu ESP8266.

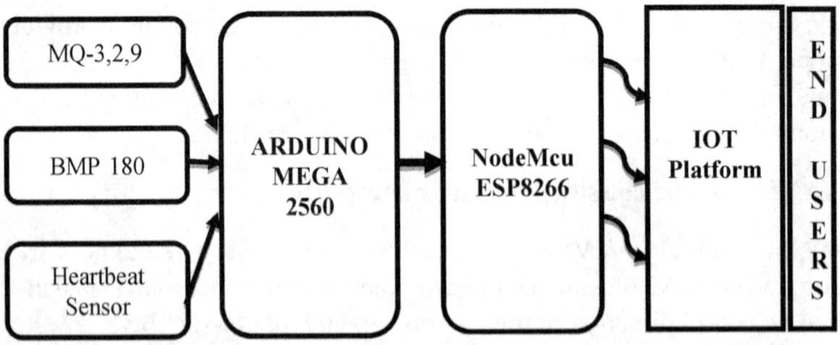

Fig. 1 Basic block diagram of the inter-vehicular information system on IOT platform

Before uploading, it was necessary to install the specific driver of NodeMcu. The code should be uploaded in NodeMcu with proper API key, username, password, and URL link so that NodeMcu can connect with the mentioned channel of desired IOT platform. There will be two types of API key-Read and Write. Write API key is used to write data to a channel and read API key is used to allow other people to view your private channel feeds and charts. For the code, write API key is needed which is case sensitive. Both username and password are user-defined. I have made this project compatible to V2S and V2P network system for the end users, so that along with roadside unit, normal people can also take part in this monitoring system (Fig. 2).

2.1 Flowchart of the Proposed System

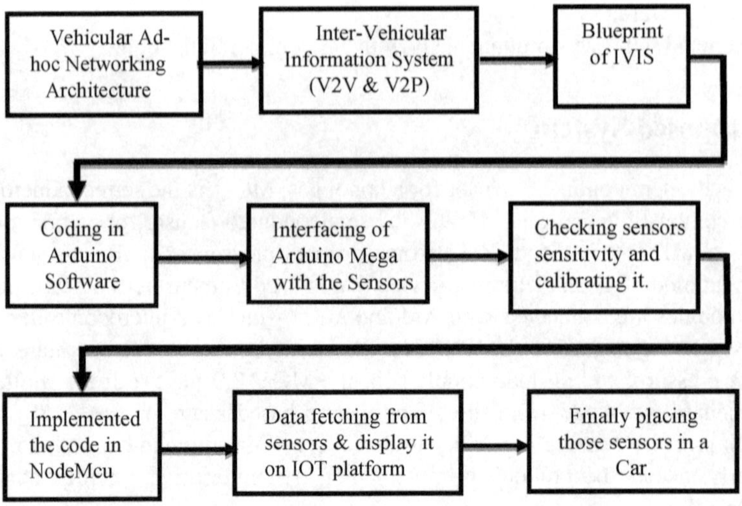

Fig. 2 Flowchart of the inter-vehicular information System (IVIS) on IOT Platform

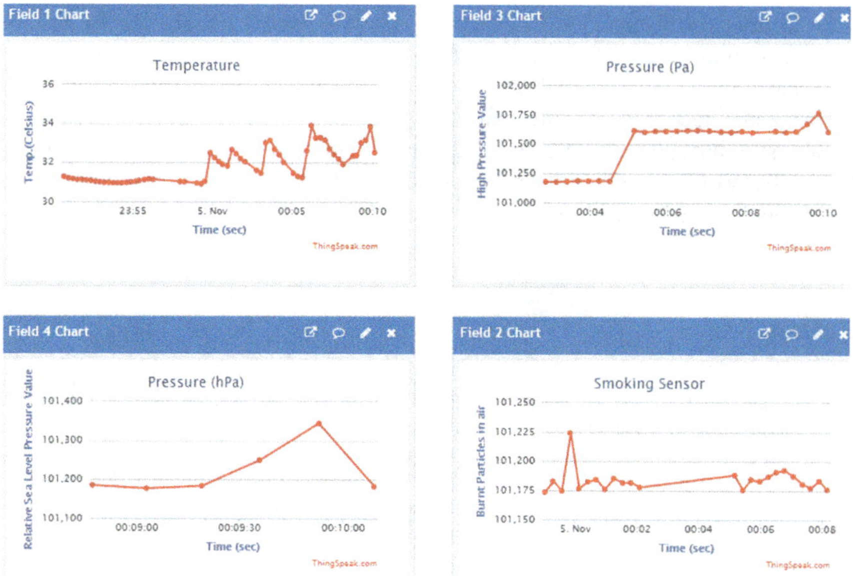

Fig. 3 Field 1, 3, and 4 charts are BMP sensor output and Field 2 chart is MQ2 sensor output

3 Results and Discussions

For an initial study, I have tested four features, i.e., temperature, absolute pressure, relative sea-level pressure, and smoke in the air. I displayed the outputs in the graphical plot in my channel of IOT platform (Fig. 3). For an initial testing purpose, I have tested the BMP sensor by exhaled air from the mouth in the room temperature. The peak point of temperature graph was achieved by increasing the encompassed temperature by burning a match stick around it. MQ2 sensor was also being tested, and the peak point was achieved from cigarette smoke. These outputs are non-calibrated; hence, calibration is required. Similarly, alcohol detection (MQ3), carbon monoxide detection (MQ9), and heartbeat rate also can be tested and plotted too.

4 Conclusion

I have used low-cost sensors, microcontroller unit, so in order to make it better, I have to use more precis sensitive sensors for industrial purposes. Although with these low-cost sensors, I have achieved my desired efficient outputs, but calibrating these values might be difficult. For a good calibration, we should have good discrete, stable, and accurate output values, which we may not get it from a low-cost sensor. IVIS design will help to keep the road safe and maintain the road ethics rules.

IVIS has several applications which are specially built for roadside unit. IVIS is capable of monitoring the physical parameters of cars and its driver simultaneously. IVIS can monitor encompassed temperature around the car, physical damage made to the car, can detect speed driving, can measure relative sea-level pressure around the car along with its altitude, and carbon monoxide emission level from the car. Simultaneously, IVIS can monitor the driver's cardiac condition, alcohol consumption, and can detect the smoke inside the car.

References

1. Diaz, R.G.: Intervehicular Information System (IVIS): the basis for a tactical information system: SAE technical paper 940982, pp. 1–16 (1994)
2. Mihail, L.: Sichitiu and Maria Kihl. IEEE Commun. Surv. Tutorials **10**(2), 88–105 (2008)
3. Toor, Y., Muhlethaler, P., Laouiti, A.: Vehicle Ad Hoc networks: applications and related technical issues. IEEE Commun. Surv. Tutorials **3**(10), 74–88 (2008)
4. Moharrum, M., Al-Daraiseh, A.: Toward secure vehicular Ad-hoc networks: a survey. IETE Tech. Rev. **1**(29), 80–89 (2012)
5. Sumayya, P.A., Shefeena, P.S.: Vanet based vehicle tracking module for safe and efficient road transportation system. **46**, 1173–1180 (2015)

A Technique for Generation of Renewable Electrical Energy from Noise

Soumyadip Sarkar, Aheli Das and Arijit Saha

Abstract In today's world, we cannot imagine only one day without using electrical or electronic gadgets. For this, electric consumption is increasing at a large rate everyday but the sources for producing electrical energy are limited. The energy sources, those are renewable, are the need for the day due to the crisis of unconventional energy resources and also the pollution caused by them. Various renewable energy resources, like wind, solar, hydro, are being used globally. Researchers are trying to identify new sources of renewable energy. In this respect, one of the unexplored forms of energy we come across is sound energy. However, sounds especially noise is a little bit unexplored renewable energy resource. Almost every product and instrument we use makes a sound. Noise is usually an unwanted sound which often creates hindrance in our desired activity. But if we can generate electricity from these various unwanted noises, then that unwanted noise can be used for a usable purpose. This present communication gives us an idea about a relatively less popular resource of unconventional energy. Through this communication, we propose to use noise for generation of electricity. Almost worthless noise can be changed to electric power and finally can be used in our various daily activities and further broader purposes.

Keywords Noise · Renewable energy · Piezoelectric

1 Introduction

Until now, the main power source throughout the world is conventional or non-renewable source which is primarily fossil fuels like coal, natural gas, crude oil. But recent estimates predict that by 2030 the fossil fuels will be completely used up, after which the world will have to rely on other energy resources. So today scientists are searching for various renewable sources of energy, which might be used to fulfil our needs in the modern growing technology-based days. Energy from sunlight

---

S. Sarkar (✉) · A. Das · A. Saha
Department of Electronics & Communication Engineering, B. P. Poddar Institute of Management & Technology, 137, VIP Road, Kolkata 700052, India
e-mail: soumyadip2302@gmail.com

J. K. Mandal et al. (eds.), *Information, Photonics and Communication*,
Lecture Notes in Networks and Systems 79,
https://doi.org/10.1007/978-981-32-9453-0_11

and wind is being used globally and is widely accepted [1–3]. Some other sources like ocean, hydropower, geothermal, biofuel are also being considered as renewable energy resources. However, sound is another unexplored source of unconventional energy. Recently, some research works are also going on in this field. Bhatnagar has mentioned about using conversion of sound into electrical energy to minimize the insufficiency of electrical energy globally and also to reduce CO_2 in the atmosphere [4]. There is a high chance of getting huge amount of electricity in nuclear power plant because huge amount of sound is produced during nuclear fission. Gupta has used piezoelectric material in conversion of electricity from sound. Here also attempt to reduce deficiency of electricity has been emphasized. Apart from that, possibility of invention of portable charging phone is also mentioned [5]. Ahmed Jamal et al. emphasized on designing a portable cell phone battery charger which is used to produce energy from sound or air pressure [6].

Sound is actually a mechanical wave that can travel through any medium. Noise is its unwanted form that means when sound intensity exceeds certain limit. For example, if sound level exceeds 85 dB approximately, then it may cause hearing problem for human being. Limitless noise is also caused other severe health issues. However, we can make fruitful use of this noise.

In this communication, we present a method to generate electrical energy making use of noise. Here we make use of piezoelectric material. It is an effective sensor. It produces small voltage corresponding to the pressure applied on it. Hence, it converts efficiently the sound vibration into electric energy. We can use this electricity in different applications. In Sect. 2, our discussion is about our proposed method fully, while the results are given in Sect. 3, followed by conclusions and future scope of research are in Sects. 4 and 5.

2 Proposed Method

According to thermodynamics, the "law of conservation of energy" says that the net amount of energy of isolated system is always constant. We know that one cannot produce or destroy energy, and energy is changeable from one to another form.

Thus, it is not impossible to get electric energy from sound. In fact, the conversion of electric energy from sound energy is not a new concept. This was first reported by the researchers from Los Alamos National Laboratory USA et al. [7]. But in that case, the sound was not naturally created to generate electricity because electricity was not created naturally from available sound and noises present in the environment. Here we propose a method to convert noise into electric energy using piezoelectric material.

The method uses piezoelectric device. In this method, the noise is changed into electric energy using piezoelectric device. When we apply strain on the surface of the crystal, and then the produced deformation is converted into electricity. In our day-to-day life, a number of noise sources lie around us like TV, mixer-grinder, generator.

If we keep the sound receiver in front of such appliances, it will convert the generated sound into electricity and can be stored in a battery. For the piezoelectric material,

$$\text{Total induced charge} = \text{Applied force} \times \text{Piezoelectric constant}$$

The entire proposed system of sound energy to electric energy conversion is illustrated by the given schematic diagram. From the figure, we find that after piezoelectric material an amplifier is used. This is required as the voltage produced by piezoelectric material corresponding to given sound intensity is relatively small. So we have to amplify this and then we use RC filter. The sound source is here used to generate small voltage varying from 150 mV to 2 V. From the filter's output, we get DC output and it is stored in battery (Fig. 1).

In few cases, we may require more number of piezoelectric sensors in order to catch the sound from all sources. The main target of this is street lightning by the use of sound of vehicles. In this case, we use a group of piezoelectric sensors. Each sensor will absorb sound, and they are located in each corner of street. The outputs of all sensors are added up by using adder circuit. Each sensor receives noise and produces small amount of voltage which we store in capacitor. Then from each capacitor voltage comes to adder circuit and added up. After adding the voltage, it is then going through RC filter and amplified. Finally, the DC output voltage is stored in battery (Fig. 2).

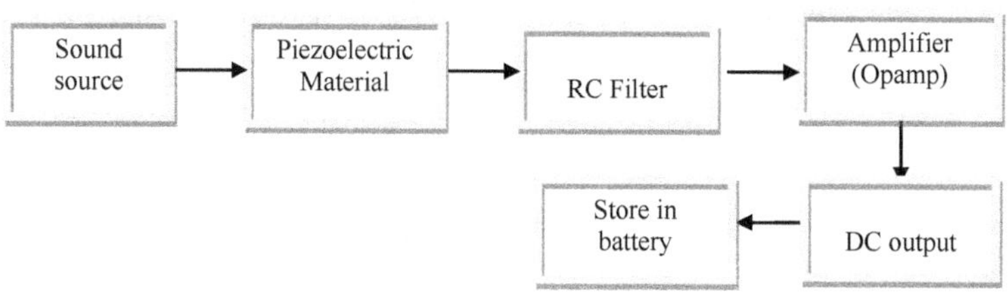

Fig. 1 Schematic diagram of conversion of sound into electric energy

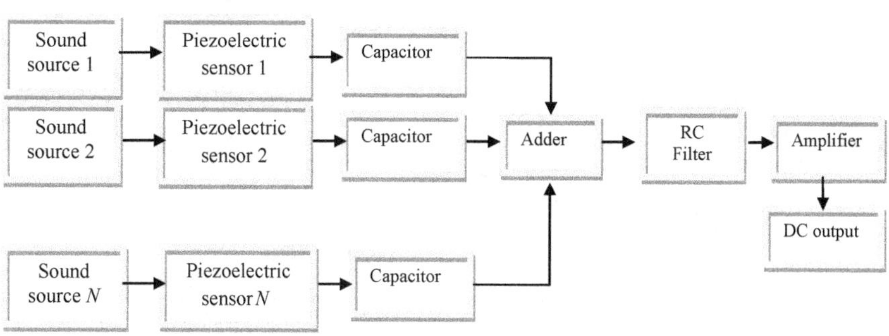

Fig. 2 Schematic diagram for generating electric energy from many sound energy sources

3 Results

With this proposed system, we have obtained different voltages for different sound intensity levels. We find that the system is generating more voltage for more sound intensity, as expected. The detailed result is provided in Table 1. From the data we obtained in Table 1, result is plotted as provided in Fig. 3.

From Fig. 3, we find that the efficiency of our proposed process is comparatively higher than the reported diaphragm technique [7] and more voltage is produced for a particular sound intensity. Though some input voltage has to be applied after conversion of sound, it produces much more voltage almost twice of initial voltage. If sound intensity gets higher, then more voltage will be produced. We know that for solid vibrating surface sound power is proportional to surface area and mean square of vibrating velocity. So for eccentric rotating loaded machines by increasing surface, we can increase sound power and from it more electricity will be produced.

4 Conclusion

There are numerous unnoticed and unpredictable sources of sound. Use of transducers in converting noise into electric energy states that the sound from the noise can act as an another energy source as well. In our present society, unpleasant sound, i.e. noise, is one of the main causes of our disturbance but on the other side the sound of

Table 1 Sound intensity vs generated voltage

Sound intensity (dB)	Voltage (V)
98.8	6.72
99.5	7.23
100.1	7.84
101.5	8.79
102.3	9.55
103.2	9.8
103.5	9.98
104	10.01
105	10.28
105.6	10.44
106	10.51
106.5	10.65
107.2	10.72
107.6	11.02
108.5	11.55
109	12

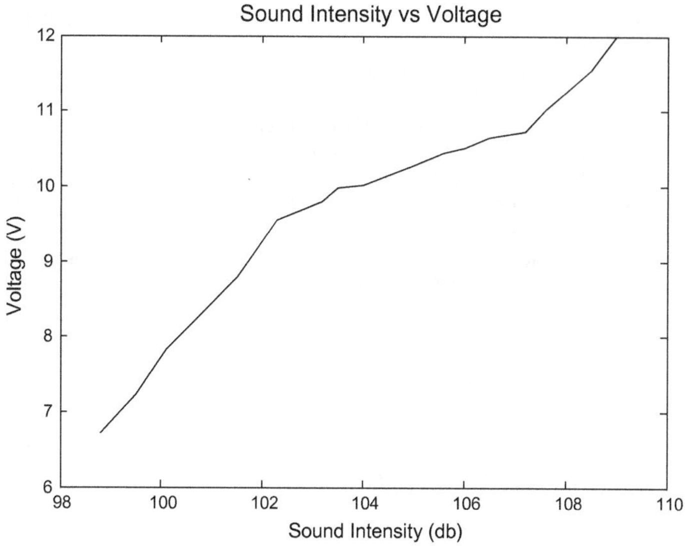

Fig. 3 Illustration of sound intensity vs generated voltage

noise can be changed to a clean energy source. Here how electricity can be generated from only one sound source as well as many random sound sources is discussed. To operate small apparatus as well as street lightning can be done by this electric energy. So our work shows a new green energy source, i.e. noise, which has a capacity to meet the requirement of electricity in our present and also future world.

5 Future Scope

This type of device can be placed in those areas where large amount of noise or sound energy is created like in nuclear power plants, motorbikes, heavy machinery, on the busy streets, stadium as this device will generate green energy. Besides these uses, we can also use this while playing electric instruments like electric guitar and so on to generate electricity from the sound produced. With the technological advancement, it is also possible for a sound absorber will absorb the loud sound from TV or generator and produce electricity which can be used for numerous purposes.

References

1. Burton, T., Jenkins, N., Sharpe, D., Bossanyi, E.: Wind Energy Handbook in Technology & Engineering. Wiley, New York (2011)
2. Ginley, D., Green, M.A., Collins, R.: Solar energy conversion toward 1 terawatt. MRS Bull. **33**, 355–364 (2008)

3. Zahedi, I.S.A.: Solar photovoltaic (PV) energy; latest developments in the building integrated and hybrid PV systems. Renew Energy
4. Shalabh Rakesh Bhatnagar: Converting sound energy to electric energy. Int. J. Emerg. Technol. Adv. Eng. **2**(10), 267–270 (2012)
5. Gupta, A., Goel, V., Yadav, V.: Conversion of sound to electric energy. Int. J. Sci. Eng. Res. **5**(1), 2146–2149 (2014)
6. Ahmed Jamal, G.R., Hassan, H., Das, A., Ferdous, J., Lisa, S.A.: A novel battery charger operated from random sound sources or air pressure. In: International Conference on Informatics, Electronics & Vision (ICIEV) (2014)
7. Mohammed Ovaiz, A., Rajkumar, S., Rajkumar, K., Ramprasad, D.: Generation of electricity from industrial noise. e-ISSN: 2395-0056, p-ISSN: 2395-0072

Signal Processing and Computational Intelligence

Mining User's Data Based on Customer's Rating for Prediction and Recommendation—A Comparative Analysis

Soma Bandyopadhyay, S. S. Thakur and Jyotsna Kumar Mandal

Abstract The business of E-commerce is increasingly becoming popular due to pervasive Internet technologies. It is a human tendency to rely on the data or information, which they receive from their friends and neighbours prior to taking any decision, especially before purchasing any item. Presently, people are getting vast information and worldwide data though Web. Due to information overload, customers often face difficulties to locate their item of interest. Recommender system plays a significant role, and it helps to deal with information overload and further provides personalized recommendations to customers or users. In this paper, recommendation of smartphone was given based on feedback given by customer using weighted mean approach. The prediction was calculated for untried items, based on ratings given by new user using collaborative filtering. The results of recommendation and prediction show the approach is interesting.

Keywords E-commerce business · Rating · Recommendation system · Collaborative filtering · Prediction

1 Introduction

In the last decade, recommender systems become an integral part of e-commerce business to promote product sell and thus become a popular research field in the present era. To provide recommendation services, online stores use mostly either collaborative filtering (CF) or content-based (CB) recommendation approach. Two types of collaborative filtering are used in the recommendation system, one is user-

S. Bandyopadhyay (✉) · S. S. Thakur
MCKV Institute of Engineering, Howrah, West-Bengal, India
e-mail: somabanmuk@yahoo.co.in

S. S. Thakur
e-mail: subroto_thakur@yahoo.com

J. K. Mandal
University of Kalyani, Nadia, West-Bengal, India
e-mail: jkm.cse@gmail.com

J. K. Mandal et al. (eds.), *Information, Photonics and Communication*,
Lecture Notes in Networks and Systems 79,
https://doi.org/10.1007/978-981-32-9453-0_12

based CF and another one is item-based CF. In the case of user-based collaborative filtering, the customers who have purchased and rated similar items with the target user are found. In the case of item-based collaborative filtering, instead of finding, similar customers focus is given on finding similar items. The algorithms attempt to find a similar type of items for each user which they have previously purchased, used or rating has been given by them [1]. The software tools and techniques that suggest the target user to choose items of their preference are known as recommender systems [2]. It can assist target user by providing a suggestion about items like products, services, information, which they would like or suits their needs [3]. Recommendation of a particular item may be provided by the following ways: items frequently purchased together, customers who purchased a particular item also purchased earlier, customers who searched a particular item also searched and sponsored product related to any particular item. According to Liang et al. [5], Web-based recommendation systems have gained popularity due to its ability of reducing overloaded information and increasing user satisfaction [4, 5]. A lot of information is needed to personalize recommender systems. User profile, age, gender, profession, educational qualification, hobby, preference, and previous purchasing pattern can be taken into consideration. The user is the most challenging to define as every user is having individual needs and goals. The loyalty of the customers should be taken into consideration. The older rating also should be considered in a good recommender system. Many recommender systems have inbuilt functions that give less weightage to older ratings but also make the system prone to possible loss of user's long-term interests that are not in frequent use [6]. If the recommended item is purchased by the user (e.g. products in an online store), then it can be considered as cost-effective but if after browsing the item, it is not purchased or the item is not found by the user it will not be considered as cost-effective.

2 Market Trends (2013–2018)

Since the advent of mobile phones, it has increasingly become an essential device for our everyday life. With the advancement of technology and in the era of miniaturization, smartphones become an essential component in our daily life. People can get multiple features like Internet, instant messenger, and e-mail in a single handheld device. They also can use this device in business or in pleasure. Table 1 shows the number of mobile phone user from 2013 to 2018 [7]. Figure 1 shows the increasing trend of the mobile phone user.

Table 2 shows the percentage of smartphone users in India from 2014 to 2018 among mobile phone users [7]. It is evident from Fig. 2 that the smartphone penetration rate is increasing in India in the recent years. The remaining paper is organized as follows. Section 3 presents the proposed work. In Sect. 4, prediction and recommendation computation have been discussed. Finally, results and discussion are presented in Sect. 5.

Table 1 Number of mobile phone user

Year	Number of mobile phone users in million
2013	524.9
2014	581.1
2015	638.4
2016	684.1
2017	730.7
2018	775.5

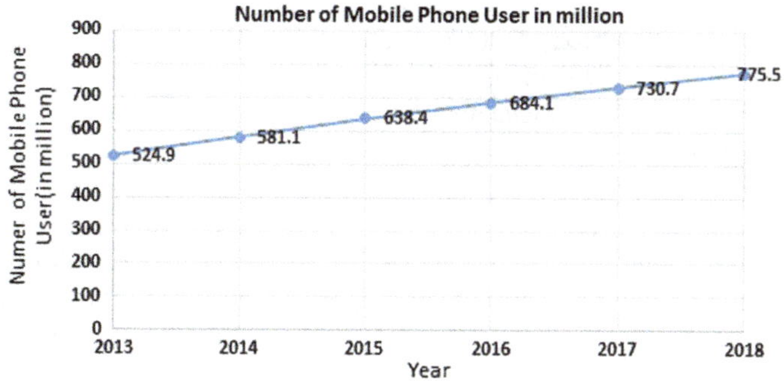

Fig. 1 Increasing trend of mobile phone user

Table 2 Smartphone user from 2014 to 2018

Year	Share of mobile phone user in %
2014	21.2
2015	26.3
2016	29.8
2017	33.4
2018	36

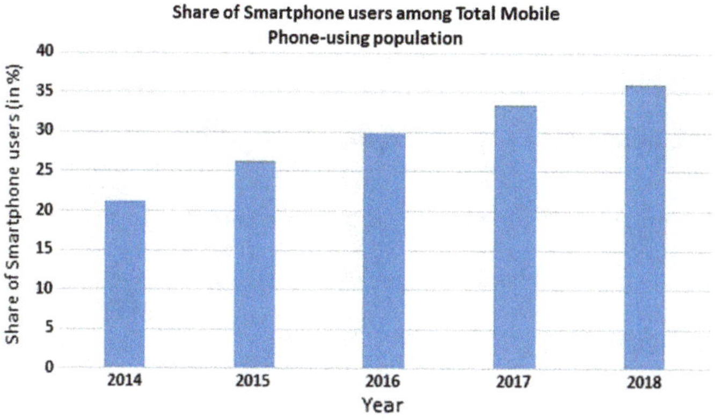

Fig. 2 Share of mobile phones users that use smartphone in India

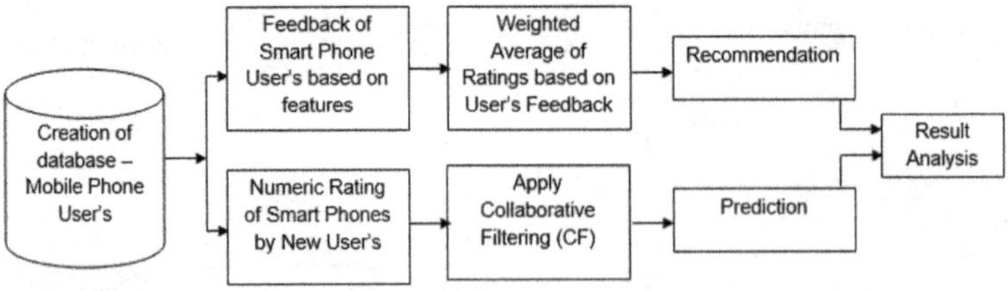

Fig. 3 Block diagram of our proposed work

A survey was conducted in 2017 among the mobile phone user. It has been observed in India that 09% of users listen to music on their mobile phones and 40% of mobile users visited social networking sites [7]. The smartphone is also used to pay utility bills. One can click a picture by using the camera of the phone or can play game using some apps. It can also be used for navigation purpose. The smartphones are not only mere a means of communication, and it can also be used for entertainment. Various companies have specifically designed smartphones targeted towards various niches. The different features on which the smartphone companies have emphasised are camera, battery, speed, display, heating issues, etc. It has been observed that people of different sex or of different age groups have different taste in mobile usage and according to their requirement, they choose the product.

3 Proposed Work

In this work, the feedback was taken from customers related to smartphones they are using at present or used in the past. The feedback was given by them based on different features of smartphones. Similarly, new customers rating [8, 9] related to smartphone were taken, and finally, a comparison was done between two results. The approach, which is presented in this section, is based on the goal of targeted marketing.

3.1 Block Diagram of Our Proposed Work

The block diagram of our proposed work has been shown in Fig. 3.

3.2 Algorithm

Step 1: Start

Step 2: Create initial database

Collect feedback from customers related to smartphones they are using at present or used in the past

Customer's feedback is based on ordinal rating

Step 3: Compute weighted average of ratings based on customer's feedback

Step 4: Consider this weighted average of ratings as user's recommendation

Step 5: Create another database

Collect feedback from new customers related to smartphones they may use in future

New customer's feedback is based on numeric rating

Step 6: Apply Collaborative filtering for prediction of items based on numeric ratings

Step 7: Do comparison between prediction and recommendation results

Step 8: Calculate accuracy and error estimation

Step 9: Repeat the process (i.e. Step 5 … Step 8) whenever a new customer gives the feedback, else go to Step 10

Step 10: Stop.

3.3 *Memory-Based Collaborative Filtering Technique*

For generating a prediction of an active user (new user), memory-based collaborative filtering algorithms use either entire database or a record of the user-item database. The people having preferences for similar kinds of products or having similar interests are treated as likeminded people. Here, neighbours of the active user (new user) are searched from the database to find the likeminded individuals and based on that prediction of preferences is produced for the new user. In case of neighbourhood-based collaborative filtering approach, first, the weight between two items is calculated. Here, $w_{i,j}$ denotes distance, correlation or weight between two users or items and i, j denote two users or items. After that, the weighted mean of all the ratings of the items are calculated for providing prediction to an active or new user searching for the similar kind of product or services [10, 11]. For generating a top-N recommendation, similarity computation (nearest neighbour) is done to find k most similar items or users. Then the neighbours are aggregated to find top-N most frequent items which will be used for recommendation.

Similarity Computation. In memory-based collaborative filtering algorithms, similarity computation between item or user is a tough task. For item-based CF algorithms, in order to find similarity, $w_{i,j}$ is calculated between two items i and j based on the users who have rated both of these items in a similar manner [10]. For user-based CF algorithm in order to find the similarity, $w_{u,v}$ is calculated between two users u and v having the same preferences of items or who have co-rated the same items. Different methodologies can be used for computing similarity or weight between items or users.

Correlation-Based Similarity. Here, similarity $w_{u,v}$ and $w_{i,j}$ are measured between two users *u* and *v*, or between two items *i* and *j*, respectively, by computing the Pearson correlation or other correlation-based similarities. To get accurate result first the co-rated cases are isolated. With the help of Pearson correlation measurement, the range to which two variables linearly relate with each other [12, 13] is computed. In case of user-based algorithm, the Pearson correlation is computed between users u and v as the following

$$w_{u,v} = \frac{\sum_{i\in I}\left(r_{u,i} - \bar{r}_u\right)\left(r_{v,i} - \bar{r}_v\right)}{\sqrt{\sum_{i\in I}\left(r_{u,i} - \bar{r}_u\right)^2}\sqrt{\sum_{i\in I}\left(r_{v,i} - \bar{r}_v\right)^2}} \tag{1}$$

where $i \in I$ summations are over the items for which both the users *u* and *v* have given rating. $\bar{r}_u$ denotes the average rating of the co-rated items of the *u*th user. Pearson correlation for item-based algorithm is calculated by Eq. 2.

$$w_{i,j} = \frac{\sum_{u\epsilon U}\left(r_{u,i} - \bar{r}_i\right)\left(r_{u,j} - \bar{r}_j\right)}{\sqrt{\sum_{u\epsilon U}\left(r_{u,i} - \bar{r}_i\right)^2}\sqrt{\sum_{u\epsilon U}\left(r_{u,j} - \bar{r}_j\right)^2}} \tag{2}$$

where the set of users who rated both items *i* and *j* are denoted by $u \in U$. In this case, $r_{u,i}$ denotes the rating of user *u* on item *i* and $\bar{r}_i$ is the average rating of the *i*th item by the same user.

Vector Cosine-Based Similarity. To measure the similarity between two vectors, vector cosine-based similarity is used. In this case, each document can be treated as a vector of word frequencies and to find similarities between two vectors the cosine of the angle which is formed by the frequency vectors is computed [14]. In the case of collaborative filtering, a similar approach is adopted where users or items are used instead of documents and ratings are used instead of word frequencies. If we consider, *R* as the $m \times n$ user-item matrix, then to find similarity between two items, *j*, the cosine of the *n*-dimensional vectors corresponding to the *i*th and *j*th column of matrix *R* and *I* are to be determined. Equation 3 shows how the cosine similarity between items *i* and *j* is computed.

$$w_{i,j} = \cos\left(\vec{i}, \vec{j}\right) = \frac{\vec{i} \cdot \vec{j}}{\left\|\vec{i}\right\| * \left\|\vec{j}\right\|} \tag{3}$$

In this work, data samples from users were collected in both online and offline modes to calculate the rating of different smartphones. Usually, the rating can be taken from a different user in the following manner:

- Numerical rating: In this case, the rating of the product can be given in the range of 1–5.

- Ordinal rating: Qualitative rating can be given by writing some comments like awesome product, average product, poor product, or comments like strongly agree, neutral, disagree.
- Binary ratings: Here, the rating can be given by writing yes or no.

Accordingly, two types of rating data were collected from the user, one is the ordinal rating based on different features of mobile which they are presently using or have an experience of use and the other is the rating on the new item. In this case, the numerical rating was taken. In the said work, the user was mainly the students and staff members of engineering colleges. Different features of the smartphone have been studied and some of the features such as display, battery backup, and hanging issue have been taken into consideration to design the questionnaire.

A survey was done among 15 students using smartphones, before the main survey to ensure that respondents understand the questionnaire. A Google form containing the questions was sent by e-mail to the targeted user. The users were requested to participate in the survey and reply through e-mail only. Some of the incomplete responses were dropped and the rest were used in this work. A total of 6490 responses were used in this work, out of which 1640 responses were from new users. For 15 different types of smartphones, eight different features were taken into consideration which is in broad sense and further divided into subcategories for rating calculation. The names of the different mobile manufacturers have been used for the sake of research work only, and we are not biased about any particular company.

Table 3 shows the feedback of different smartphones based on their features, namely camera, battery backup, performance, and heating issues. In actual, there are 17 features, namely awesome camera, decent camera, poor camera, awesome battery backup, average battery backup, poor battery backup, awesome display, average display, poor display, awesome performance, poor performance, heating issue, and hanging issue. Due to the space constraints in the table, all the features cannot be shown but in actual while doing the calculations for recommendation all the features were taken into consideration. In this case, the user has given their remarks in different words like awesome camera, decent camera, poor camera, bad battery backup, value for money, hanging issues, heating issues, and poor performance These types of feedback were divided into 2–3 groups: good, average, and poor and according to the feedback the weighted sum was taken and rating was calculated.

Table 4 shows the user and smartphone-based rating. Actually, the rating has been taken from 1640 new users, but for the sake of simplicity, only the rating of user1, user2, and user1640 is shown in Table 4. In the first case, the rating was calculated based on the weighted average, and according to that, the recommendation was done and in the second cases, the rating was calculated based on collaborative filtering for prediction.

Table 3 Feedback of different user based on features of smartphone

Different smart-phones	Awesome camera	Decent camera	Awesome battery backup	Average battery backup	Awesome perfor-mance	Poor perfor-mance	Heating issues
Lenovo K8 Note	382	55	290	34	240	57	1240
Lenovo K8 Plus	92	14	89	6	45	39	213
Lenovo K6 Note	37	0	44	2	16	13	79
Lenovo A7000	185	27	259	14	134	46	765
Samsung Galaxy A6	68	7	20	3	14	0	6
Moto G6	67	34	44	17	16	36	0
Moto G5 S Plus	26	1	17	1	13	2	40
Realme 1	47	7	35	8	28	3	63
Honor 9 Lite	113	23	40	18	40	41	214
Nokia 6.1	10	10	10	12	27	0	18
Vivo Y71	12	0	4	0	4	0	11
Xiaomi Redmi5	13	0	8	0	6	0	14
Xiaomi Redmi Note 5	94	45	98	17	39	24	126
Honor C	23	68	24	43	2	13	19
Samsung Galaxy J7 Max	169	12	174	12	89	14	34

After that, the comparison was done based on the calculation. As the data were taken from the students and staff community, so it was possible to address the people of different age groups as well as different income groups. It is a very common practice of the customer that they want to know others opinion while purchasing the same product [15].

Table 4 New user rating of smartphone

Different smartphones	User1	User2	…	User1640
Lenovo K8 Note				3
Lenovo K8 Plus				
Lenovo K6 Note	3			
Lenovo A7000	1	2		2
Samsung Galaxy A6	4.5			
Moto G6				
Moto G5 S Plus	2			
Realme 1				
Honor 9 Lite		1		
Nokia 6.1				
Vivo Y71				
Xiaomi Redmi5		4		
Xiaomi Redmi Note 5	4	5		4
Honor C		1		
Samsung Galaxy J7 Max	5			5

4 Prediction and Recommendation Computation

Collaborating filtering system is one of the important processes for obtaining predictions or recommendations. In the neighbourhood-based collaborative filtering approach, a subset of nearest neighbours of the active user are chosen by finding their similarity with the active user. Then a weighted aggregate of their ratings is computed which is used for generating predictions for the concerned user [16, 17].

4.1 Weighted Sum of Others' Ratings

The formula for finding out the prediction for the active user, *a*, on a certain item, *i* is given below.

$$P_{a,i} = \overline{r_a} + \frac{\sum_{u \epsilon U} \left(r_{u,i} - \overline{r_u} \right) w_{a,u}}{\sum_{u \epsilon U} \left| w_{a,u} \right|} \tag{4}$$

where $\bar{r}_a$ and $\bar{r}_u$ denote the average ratings for the user a and user u, respectively, on all other rated items, and $w_{a,u}$ indicates the weight between the user a and user u. The summations are over all the users $u \in U$ who have rated the item i.

4.2 Simply Weighted Average

Simple weighted average may be used for item-based prediction to predict the rating, $P_{u,i}$ for user u on item i [10].

$$P_{u,i} = \frac{\sum_{n \in N} r_{u,n} w_{i,n}}{\sum_{n \in N} |w_{i,n}|} \quad (5)$$

where the summations are over all other rated items $n \in N$ for user u. Here, $w_{i,n}$ is the weight between items i and n, $r_{u,n}$ denotes the rating for user u on item n.

5 Result and Discussion

The present work compares the rating system for a recommendation in two ways. While considering the rating of a particular item, attention should be given on the following things:

- Number of ratings: Number of ratings available is one of the most important data for a recommendation as this information is useful to numerous users who are searching for these items. If a user rates more items, many users can get suggestions from his or her rating.
- Degree of agreement with others: The degree of agreement is measured by comparing one user's opinion with others.
- Standard deviation of rating: It is the deviation of user's rating with his average rating. If the deviation is more, that will be given priority.
- Degree of similarity with top neighbours: The user who has similar weight can be replaced as he or she is very similar to other users while the user having greater weight will be considered as influential.
- Aggregated popularity of the rated items: The user has greater probability of having overlapped preferences when the sum of the popularities of the chosen rated items are high [13].

One can find desirable items if accurate recommendation is provided [18]. Table 5 shows the average rating (rating provided by a user on untried item) and average rating (computed based on feedback after use) of smartphones. Figure 4 shows acomparison between the rating calculations.

From the graph, it is clear that the predicted rating (rating for untried item) and calculated rating based on user feedback are closer to each other. The results of recommendation and prediction show the approach has relevance as the work is concerned and the manner the data has been dealt in for future recommendations.

Table 5 Comparison between recommendation and prediction

Smart phone	Recommendation of rating based on user feedback	Prediction of rating given by new user
Lenovo K8 Note	3.33	3.70
Lenovo K8 Plus	3.37	3.50
Lenovo K6 Note	3.58	3.50
Lenovo A7000	3.30	3.00
Samsung Galaxy J6	4.83	4.50
Moto G6	3.70	3.80
Moto G5 S Plus	4.09	4.00
Realme 1	4.11	3.00
Honor 9 Lite	3.25	3.00
Nokia 6.1	3.59	3.50
Vivo Y71	3.52	3.75
Xiaomi Redmi5	4.36	4.00
Xiaomi Redmi Note 5	4.00	4.00
Honor C	3.91	3.60
Samsung Galaxy J7 Max	4.49	4.50

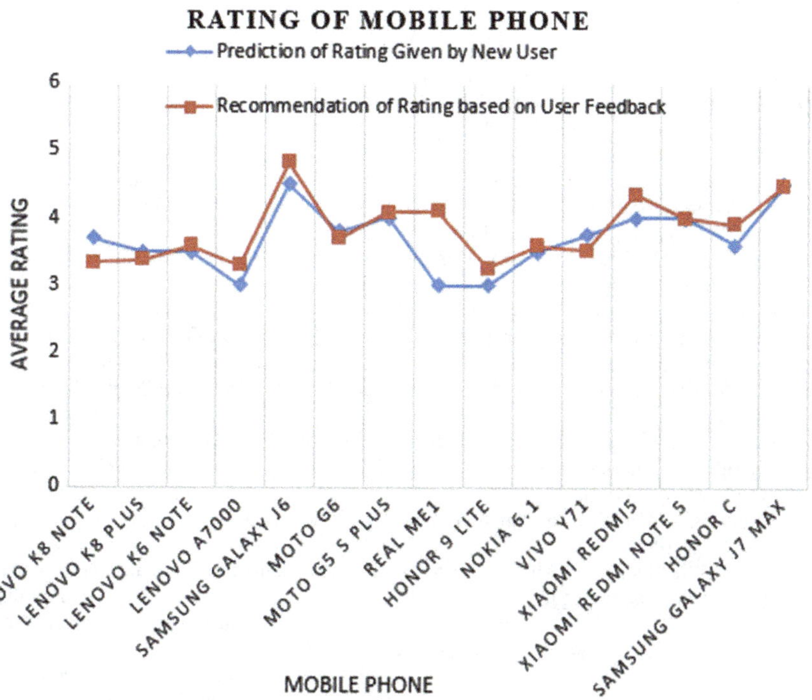

Fig. 4 Comparison between the rating calculations (recommendation vs. prediction)

Acknowledgements The authors are thankful to Director, MCKVIE and Principal, MCKVIE, for providing the computer laboratories and other infrastructure to do the proposed work. The authors are also thankful to the students of CSE department of MCKVIE for collecting the required data for the proposed work.

References

1. Linden, G., Smith, B., York, J.: Amazon.com recommendations: item-to-item collaborative filtering. IEEE Internet Comput. **7**(1), 76–80 (2003)
2. Rokach, L., Ricci, F., Shapira, B., & Kantor, P.B.: Recommender Systems Handbook. Springer, Berlin (2011)
3. Mahmood, T., & Ricci, F.: Improving recommender systems with adaptive conversational strategies. In: Proceedings of the 20th ACM Conference on Hypertext and Hypermedia, pp. 73–82, Torino, Italy (2009)
4. Liang, T.P., Shaw, J.P., Wei, C.P.: A framework for managing web information: current research and future direction. In: Proceedings of the 32nd Hawaii International Conference on System Sciences (1999)
5. Liang, T.P., Hu, P.J.H., Kuo, Y.R., Chen, D.N.: A web-based recommendation system for mobile phone selection. In: 11th Pacific-Asia Conference on Information Systems, Auckland, New Zealand (2007)
6. Burke, R.: Hybrid web recommender systems. In: The Adaptive Web, pp. 377–408. Springer, Berlin (2007)
7. http://www.statista.com
8. Dhawan, S., Singh, K., Jyoti: High rating recent preferences based recommendation system. Procedia Comput. Sci. **70**, 259–264 (2015)
9. Dhawan, S., Singh, K., Kumar, N.: Rating based mechanism for effective ecommerce product recommendation in social networks. J. Netw. Commun. Emerg. Technol. (JNCET) **7**(9), 18–21 (2017)
10. Sarwar, B.M., Karypis, G., Konstan, J.A., Riedl, J.: Item based collaborative filtering recommendation algorithms. In: Proceedings of the 10th International Conference on World Wide Web (WWW '01), pp. 285–295. Hongkong (2001)
11. Xie, F., Xu, M., Chen, Z.: RBRA: a simple and efficient rating-based recommender algorithm to cope with sparsity in recommender systems. In: 26th International Conference on Advanced Information Networking and Applications Workshops, pp. 306–311 (2012)
12. Resnick, P., Iacovou, N., Suchak, M., Bergstrom, P., Riedl, J.: GroupLens: an open architecture for collaborative filtering of netnews. In: Proceedings of the 1994 ACM Conference on Computer Supported Cooperative Work, pp. 175–186. Chapel Hill, North Carolina, USA (1994)
13. Rashid, M., Karypis, G., Riedl, J.: Influence in ratings-based recommender systems: an algorithm-independent approach. In: The SIAM International Data Mining Conference, pp. 57–66. Newport Beach (2005)
14. Salton, G., McGill, M.: Introduction to Modern Information Retrieval, p. 124. McGraw-Hill, New York (1983)
15. Sohail, S.S., Siddiqui, J., Ali, R.: User feedback based evaluation of a product recommendation system using rank aggregation method. In: El-Alfy, E.S., Thampim S., Takagim H., Piramuthum S., Hanne, T. (eds.) Advances in Intelligent Informatics. Advances in Intelligent Systems and Computing, 320, pp. 349–358. Springer, Cham (2015)
16. Herlocker, J.L., Konstan, J.A., Borchers, A., Riedl, J.: An algorithmic framework for performing collaborative filtering. In: Proceedings of the 22nd Annual International ACM SIGIR Conference on Research and Development in Information Retrieval (SIGIR '99), pp. 230–237. Berkeley, California, USA (1999)

17. Su, X., Khoshgoftaar, T.M.: A survey of collaborative filtering techniques. Adv. Artif. Intell. **2009**, 1–19 (2009)
18. Yang, X., Guo, Y., Liu, Y., Steck, H.: A survey of collaborative filtering based social recommender systems. Comput. Commun. **41**, 1–10 (2014)

Charlier and Meixner Moments and Their Application for Texture and Image De-noising Problems

Perugu Ananth Raj

Abstract Moment-based features have been employed for solving various problems in pattern recognition and image processing. In this paper, we choose Charlier and Meixner moments separately, to solve texture segmentation, classification, and de-noising of images corrupted with Gaussian noise. In order to solve the texture classification problem, we propose to use lower-order Charlier and Meixner moments of 12 Haralick's texture features and a two-class support vector machine. But, for texture segmentation problem, we use these moment-based texture energy features suggested for geometric moments by Mihran Tuceryan. Further, these energy features are used in *K*-means, Fuzzy *K*-means, and Kohonen's neural network for solving texture segmentation problem, whereas for image de-noising problem, we use non-local mean (NLM) filter with Charlier and Meixner moments-based similarity values instead of pixel-based similarity values used in NLM for filter weights calculation. The simulation results show that the proposed applications worked well for two-class texture segmentation, classification of texture images and de-noising of images corrupted with a small amount of noise.

Keywords Charlier and Meixner moments · Gray-level co-occurrence matrix · Non-local mean filter

1 Introduction

Problems either in image processing or in pattern recognition require image representation by a set of values which are invariant to various distortions. Even though there are many methods available in the literature, orthogonal moment-based image representation gained popularity because of its good image representation capability under different distortions and less sensitivity to noise [1]. Hence, they are used as features for solving different problems in image processing and pattern recognition. Some of the well-known orthogonal moments used for image represen-

P. A. Raj (✉)
Department of ECE, College of Engineering, Osmania University, Hyderabad 500007, India
e-mail: ananthraj0123@gmail.com

J. K. Mandal et al. (eds.), *Information, Photonics and Communication*,
Lecture Notes in Networks and Systems 79,
https://doi.org/10.1007/978-981-32-9453-0_13

tation are Geometric, Legendre, Zernike, Tchebichef, Krawtchouk Gaussian Hermite, Charlier, Meixner, etc. Most of the work on Charlier and Meixner moments was connected with image analysis and image reconstruction [2–4] tasks only. Till now, no attempt has been made to use them for texture classification, segmentation, and de-noising of images corrupted with Gaussian noise. This motivated us to use these two moments for solving these problems. One of the earliest works on texture segmentation using lower-order geometric moments was reported by Tuceryan [5], after that, many research papers were published on texture classification using Tchebichef moments [6]. Recently, Di Ruberto et al. [7] reported another moment-based method for texture classification that uses Legendre and Tchebichef moments of the gray-level co-occurrence matrix and obtained results and was compared with a convolutional neural network. Gelzinis et al. [8] also reported another work on texture classification. In their work, 14 Haralick's texture features [9] at different distances were fitted with a *n*th degree polynomial, parameters of the polynomial were used for texture features. For example, a total number of features for a texture image is (order of the polynomial + 1) * 14. An excellent review paper on texture representation and segmentation work carried out since 2000 was reported by Liu et al. [10]. In our work, instead of fitting with a polynomial for better discrimination of Haralick's features, we compute Charlier and Meixner moments of 12 Haralick's features. The gray-level co-occurrence matrix used for computing 12 Haralick's features was obtained by taking the average of four GLC matrices, computed at a unit distance and $0, 45, 90$, and 135° orientations. This action of taking moments of Haralick's features provides a better discriminability as well as size reduction, because in our work, we use only second-order moments for texture feature representation.

Next problem, we have attempted was texture segmentation using these two moments. The proposed approach is an extension of Tuceryan's [5] and Ananth Raj and Venkataramana's [11] work for Charlier and Meixner's moments. In this work, lower-order Charlier and Meixner moments were computed in a small window of the image and transformed them into texture energy features using the tanh function after subtracting mean from it. This step is required to capture higher-order features like termination, closure, and linearity present in the texture images. These features are submitted to *K*-means, Fuzzy *K*-means, and Kohonen's neural network for texture segmentation task. Another problem considered was filtering using the non-local mean (NLM) filter, which filters an image using local neighborhood patches. Earlier two filters were suggested using Zernike [12, 13] and Krawtchouk [14] moments for Cryo-EM images and SAR images. Zernike moments are orthogonal over a unit circle. Hence, we get errors when they were calculated over a square image region. In our approach, pixel weights would be calculated using Charlier and Meixner moments instead of pixel values used in the standard filter.

This paper is organized as follows: In Sect. 2, we give necessary details on Charlier and Meixner moments. In Sect. 3, we present the problem statement and methodology used for solving the problem. Finally, Sects. 4 and 5 give simulation results, conclusions, and extensions about the work.

2 Discrete Orthogonal Moments

In general, discrete orthogonal moments of a signal are obtained by projecting a signal onto a polynomial kernel function. In this work, we use both Charlier and Meixner kernel functions to define these moments. Hence, a brief overview is given in this paper. These orthogonal polynomials are normalized to avoid fluctuations in the numerical calculations. Figure 1 shows a normalized Charlier and Meixner polynomials for $N = 256$ points and order 10 obtained using the three-term recurrence equation [1] in terms of x.

Recently, Di Ruberto et al. [7] used Stieltjes method to generate Legendre and Chebyshev polynomials and applied them for texture analysis and image reconstruction problems. Modified Chebyshev algorithm [15] is another method that can be used for generating orthogonal polynomials. This method is not that popular, because selection of initial moments is a difficult task. Charlier polynomials are orthogonal with respect to the Poisson distribution on the nonnegative integers [15], and Mexiner polynomials are also defined on the lattice N and orthogonal with respect to the negative binomial distribution [15]. One dimensional normalized Charlier $\check{C}(p, a, x)$ and Meixner $\check{M}(p, \beta, \mu, x)$ polynomials of order p are defined [16, 17] by

$$\check{C}(p, a, x) = C(p, a, x)\sqrt{\frac{w(x)}{\rho(p)}}$$

where $\rho(p) = \frac{p!}{a^p}$, $w(x) = \frac{e^{-a}a^p}{x!}$, a is a positive constant.

$$w(x+1) = \frac{a}{x+1}w(x) \quad w(0) = e^{-a}$$

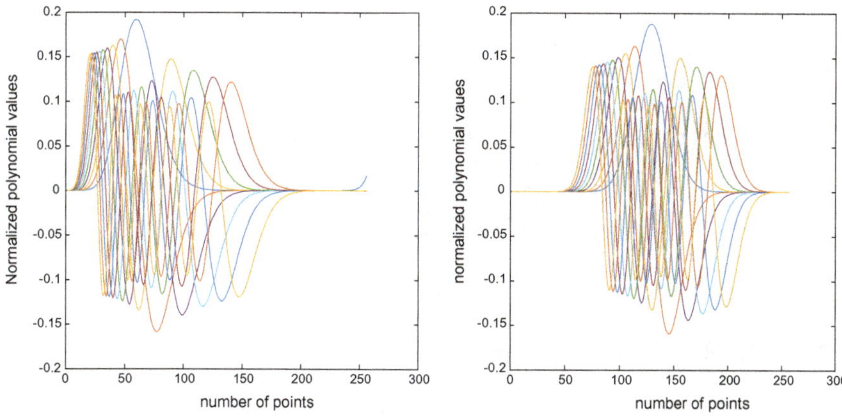

Fig. 1 Plot of normalized Meixner and Charlier polynomials of order 10, $N = 256$ points, x-axis in both cases denote no. of points

$$\rho(p+1) = \frac{p+1}{a}\rho(p)$$

$$\check{M}(p, \beta, \mu, x) = M(p, \beta, \mu, x)\sqrt{\frac{w(x)}{\rho(p)}}$$

$$\rho(p+1) = \frac{(p+1)(\beta+p)}{\mu}\rho(p) \qquad \rho(0) = \frac{1}{(1-\mu)^{\beta}}$$

$$w(x+1) = \frac{\mu(\beta+x)}{(x+1)}w(x), \quad \mathrm{w}(0) = 1$$

One of the important properties of these two polynomials is the availability of three-term recurrence equation in terms of x and order n. The first equation is used for forward moment computation and the second equation is used for inverse moment transform. Selected applications in this paper require only forward moments. Hence, we give expressions for forward moment computation only. Charlier moments of order $(p+q)$ for an image f of size $N \times N$ pixels are

$$C_{pq} = \sum_{x=0}^{N-1}\sum_{y=0}^{N-1} f(x, y)\check{C}(x, p, a)\check{C}(y, q, a) \tag{1}$$

Expression for Meixner moments M_{pq} of order $(p+q)$ for an image f of size $N \times N$ pixels can be obtained by replacing the above Charlier moments term $\check{C}(.)$ with Meixner moments $\check{M}(\cdot)$ term. A three-term recurrence formula is generally used for generating normalized Charlier polynomials with respect to x are given by

$$\check{C}(p, a, x) = \frac{\sqrt{w(x)}}{\sigma(x-1)+\tau(x-1)}\left[\frac{2\sigma(x-1)+\tau(x-1)-\lambda_n}{\sqrt{w(x-1)}}\check{C}(p, a, x-1) - \frac{\sigma(x-1)}{\sqrt{w(x-2)}}\check{C}(p, a, x-2)\right] \tag{2}$$

Initial values required for solving the above equation are

$$\dot{C}(p, a, 0) = \sqrt{\frac{w(0)}{\rho(p)}}\dot{C}(p, a, 1) = \frac{a-p}{a}\sqrt{\frac{w(1)}{w(0)}}\dot{C}(p, a, 0),$$

$$\sigma(x) = x, \quad \tau(x) = a - x, \quad \lambda_n = n$$

Normalized Meixner polynomials with respect to x is

$$\check{M}(p, \beta, \mu, x) = \frac{\sqrt{w(x)}}{\sigma(x-1)+\tau(x-1)}\left[\frac{2\sigma(x-1)+\tau(x-1)-\lambda_n}{\sqrt{w(x-1)}}\check{C}(p, \beta, \mu, x-1) - \frac{\sigma(x-1)}{\sqrt{w(x-2)}}\check{C}(p, \beta, \mu, x-2)\right] \tag{3}$$

Initial values required for solving the above equation are

$$\check{M}(p,\beta,\mu,0) = (\beta)_p\sqrt{\frac{w(0)}{d_n^2}}\cdot\check{M}(p,\beta,\mu,1)$$
$$= \frac{\mu(p+\beta)-p}{\mu\beta}\sqrt{\frac{w(1)}{w(0)}}\check{M}(p,\beta,\mu,0)$$

$$(\beta)_0 = 1(\beta)_k = \beta(\beta+1)(\beta+2)\ldots(\beta+k-1),\quad k=1,2,3\ldots d_p^2$$
$$= \frac{p!(\beta)_p}{\mu^p(1-\mu)^\beta}$$

$$\tau(x) = \beta\mu - x(1-\mu),\quad \lambda_p = p(1-\mu),\quad \sigma(x) = x$$

In this work, moment order equal to $(p+q)=3$ and a small image patch of size 7×7 at each image pixel position was used for texture segmentation problem, but for texture classification problem, we compute moments of 12 Haralick texture features of order 5. In the next section, we present application of these moments for texture and image de-noising problems.

3 Problem Statement and Methodology

The primary goal of this paper is to use orthogonal moment-based features to solve, texture classification, segmentation, and image de-noising problems. Problem statement and methodology for each problem is given below.

(a) Given a set of texture images, containing two classes, the task is to design a classifier that classifies correctly the trained images using orthogonal moments by reducing the feature vector size. (b) Given an image containing different types of texture, the goal is to develop an algorithm to segment the image based on features extracted using orthogonal moments. (c) Given a noisy image, the aim is to design a filter that removes noise, using NLM filter and orthogonal moments. Next, we present the methodology employed for solving the above problems.

Either the texture classification or the texture segmentation task requires two main steps: feature extraction and classification. In this paper, we first compute 12 Haralick's features [9]: Angular second moment, contrast feature, entropy feature, variance feature, correlation feature, inverse difference moment, sum average feature, sum variance feature, sum entropy feature, difference variance feature, difference entropy feature, and information measure of correlation features are computed using the gray-level co-occurrence matrix, that can be obtained by taking the average of four co-occurrence matrices computed at angles 0, 45, 90, and 135° by keeping the pixel displacement equal to unity. The overall scheme employed for texture classification is shown in Fig. 2. Let this gray-level co-occurrence matrix be denoted by G and it is calculated by

$$G = \frac{[g(1, 0°) + g(1, 45°) + g(1, 90°) + g(1, 135°)]}{4} \quad (4)$$

In the above expression $g(1, x)$ denotes the gray-level co-occurrence matrix calculated at different ($x = 0, 45, 90, 135$) angles by keeping the distance between the pixels equal to unity, using this matrix, we compute 12 Charlier and Meixner moments of 12 Haralick's feature using Eq. 5 for better discriminability

$$\check{C}_p = \sum_{x=0}^{11} F(x)\check{C}(x, p, a) \quad (5)$$

$F(x), x = 0, 1, 2, \ldots 11$, denote Haralick's 12 feature values, p denotes the moment order, for texture classification p value is set as 7. These feature values are used for training a two-class support vector machine. The same procedure was used for Meixner moments. Whereas for texture segmentation problem, we compute texture energy features using 2D Charlier moments and the procedure is the same for Meixner moments.

$$C_{pq} = \sum_{x=0}^{6}\sum_{y=0}^{6} f(x, y)\check{C}(x, p, a)\check{C}(y, q, a), \quad \text{p} = 0, 1, 2, \quad \text{q} = 0, 1, 2$$

We get a total of nine moments, but we use only six moments $C_{00}, C_{01}, C_{02}, C_{10}, C_{11}$ and C_{12} for texture segmentation task. The first value was not used because it will not contribute much information. These values are transformed into texture energy features using Eq. 6

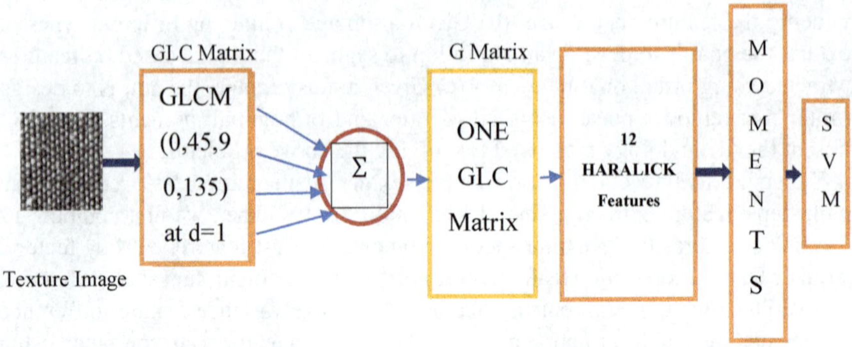

Fig. 2 Texture classification scheme

$$T_k(x,y) = \frac{1}{L^2} \sum_{(a,b)\epsilon W_{xy}} \left|\tanh\left(\sigma\left(M_k(a,b) - \check{M}_k\right)\right)\right|, k = 1, 2 \ldots 6 \quad (6)$$

where W_{xy} is a $L \times L$ size subimage window size at each pixel location, constant σ controls the shape of the logistic function. In this work $\sigma = 0.01$ was fixed, $\check{M}_k$ denoted mean of the kth feature image. With these values, every texture image will have six features at each pixel location. Hence, the total size of the data for each image is $N \times N \times 6$. These values are used for segmentation. Next, we consider the problem of image de-noising. The mathematical expression for a noisy image is $S(x, y) = f(x, y) + \mathbf{N}(x, y)$ in this expression $S(.,.)$ and $f(.,.)$ represents noisy original images, whose size is $N \times N$ pixels, $\{x, y = 0, 1, \ldots N - 1\}$, $N(.,.)$ denotes the Gaussian Noise, with mean zero and variance σ^2. In this work, we use a non-local mean filter [18] with feature-based similarity measure to process the noisy image. In traditional non-local mean filter, every noisy pixel is replaced by a weighted average (mean) of all the pixels within a pre-selected window, where the weights are computed using the neighborhood similarity of the image patches. Filter weight between any two pixel position say x, and y and processed pixel using NLM filter is given by

$$w(x,y) = \exp\left(-\frac{[[P_x - P_y]]^2}{h^2}\right) \quad (7)$$

$$\tilde{S}(x,y) = \frac{\sum_{m=-a}^{a} \sum_{n=-b}^{b} w(m,n) f(x+m, y+n)}{\sum_{m=-a}^{a} \sum_{n=-b}^{b} w(m,n)} \quad (8)$$

for $x, y = 0, 1, \ldots N - 1$.

where P_x, P_y denotes image patches at x and y, respectively. Parameter h acts as a degree of filtering, and it controls the decay of the exponential function. In our work, these two values are replaced with Charlier and Meixner moment features. The sequence of steps required for filtering a noisy image using Charlier and Meixner moment features are

Step 1: Compute Charlier and Meixner moment at each pixel position of a given noisy image using 7×7 image patch.
Step 2: Choose a similarity window of size 21×21 around each pixel position that is to be processed and compute the similarity weight between the pixel under consideration and all other pixel positions within a window of size 21×21 using Eq. 7.
Step 3: Substitute the computed weight in Eq. 8 to obtain a filtered image.

In the next section, the simulation results would be presented.

4 Simulation Results

In order to test the proposed approaches for texture classification, segmentation and de-noising of images corrupted with Gaussian, we have performed some simulations using texture images and standard bird image. We have created a two-class and four-class texture images using Brodatz texture data set downloaded from [19] for the texture segmentation problem, these images are shown in Fig. 3, but for texture classification task, we have used only two classes, each class containing five images only. These images are shown in Fig. 3a, b.

These images are also downloaded from [20], and 12 Haralick's texture features were computed using the gray-level co-occurrence matrix given in Eq. 4. Next, Charlier and Meixner moments of these 12 features are computed. These values are used in SVM classifier function which is available in MATLAB 2017 a, tool box. When an unknown data is submitted to a trained classifier using Charlier and Meixner moment features, we obtained correct results in the both cases. Different parameters used for this simulation work are (a) Charlier moments $a = 10$, order $p = 5$ and (b) Meixner moments $\mu = 0.5, \beta = 20, p = 5$, SVM classifier was trained using radial basis function. Texture images used for classification are shown in Fig. 4a, b. For texture segmentation problem, we compute Charlier and Meixner moments at each pixel position using a 7×7 subimage, except for the first moment, all other eight moments are utilized in *K*-means, Fuzzy *K*-means, and Kohonen's network for segmentation task. The number of training epochs for Kohonen's network was fixed at 3. The initial size of the neighborhood is $[3 \times 3]$, and learning rate parameter was 0.9 for Kohonen's network. These functions are also available in MATLAB tool box. Various parameters used for Charlier and Meixner moments are $a = 4.6$, order $= 2$, $\mu = 0.5, \beta = 4, p = 2$.. Results are shown in Figs. 5 and 6, respectively.

The next simulation carried out was on filtering an image corrupted with Gaussian noise. First, a bird image was corrupted with Gaussian noise with sigma $= 0.02$, both original and noisy images are shown in Fig. 7. We have fixed a search window of

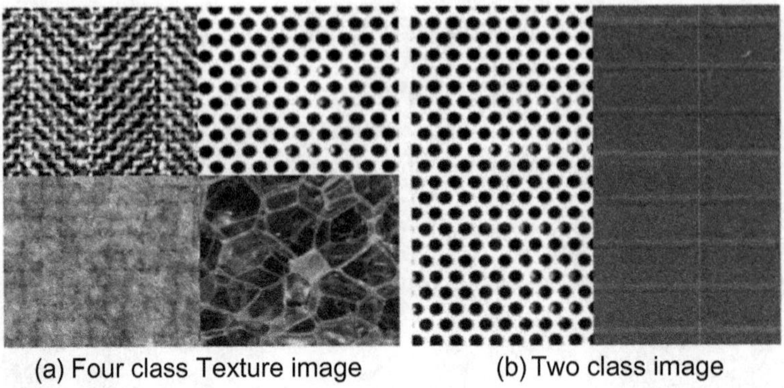

Fig. 3 Four-class and two-class Brodatz texture images

Fig. 4 **a** Image data set used for class 1. **b** Image data set used for class 2

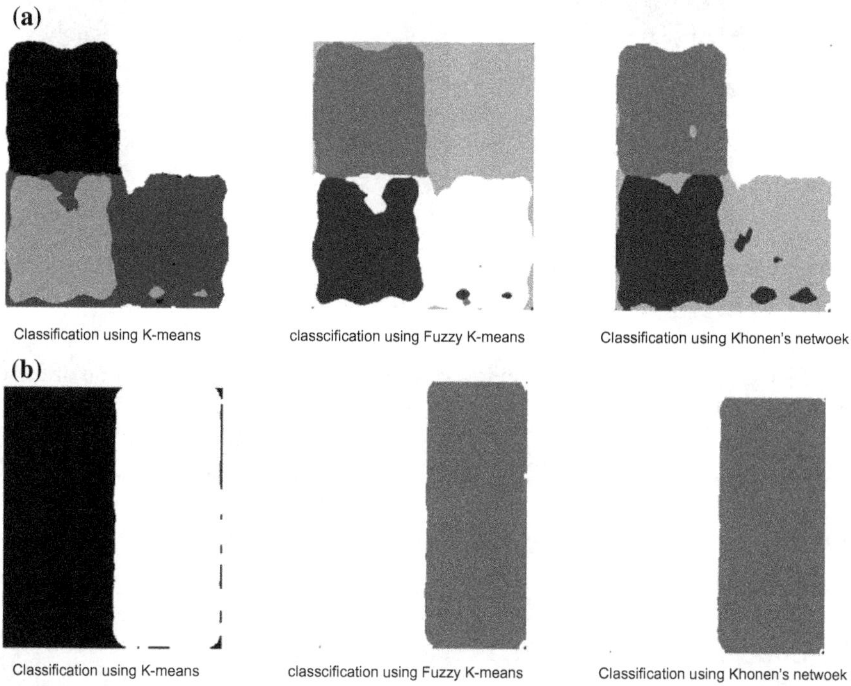

Fig. 5 **a** Four-class texture classification using Meixner moments **b** Two-class texture classification using Meixner moments

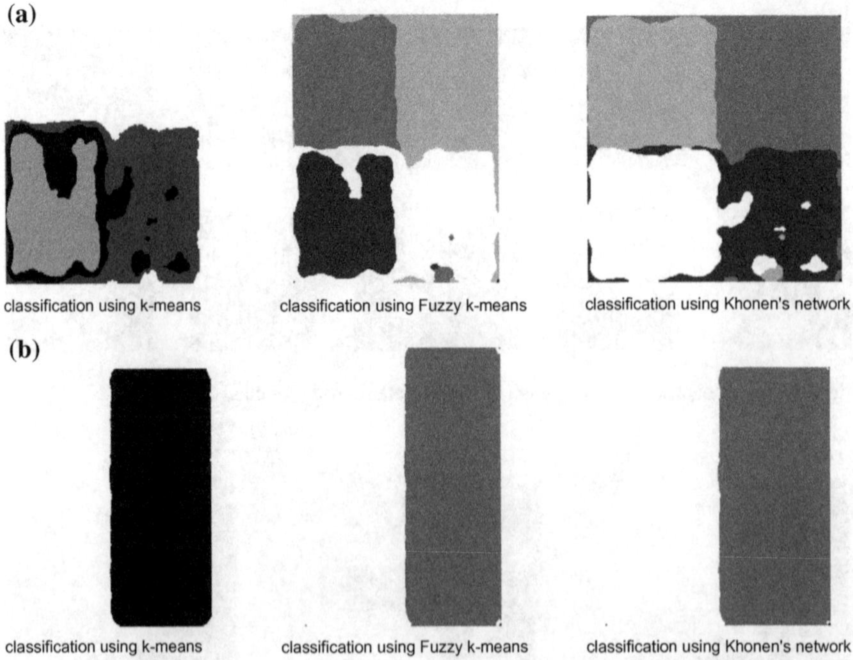

Fig. 6 **a** Four-class texture classification using Charlier moments **b** Two-class texture classification using Meixner moments

size 21 × 21 pixels and a local window of size 7 × 7 pixels for computation of these moments. We have used only eight moments for computation of filter weights and weight decay parameter $h = 0.75 * 0.08$ was used in our simulation. It is observed that when the h value is small "NO" filtering is noticed, whereas for a large value of h, we observed blurring on the image. Hence, we have selected that value of h for which no blurring occurred. Filtered images are shown in Fig. 7.

5 Conclusions

In this paper, we have used both Charlier and Meixner moments-based features for texture classification, segmentation, and de-noising of images. In order to reduce the feature size as well as better separability of Haralick's features, we use Charlier and Meixner moments of 12 Haralick's features that are computed using a gray-level co-occurrence matrix. These features are used for a two-class SVM classifier. We have also employed these two moment-based texture energy features for texture segmentation using K-means, Fuzzy K-means, and Kohonen's neural network. Further, we have also proposed a moment-based filter for processing an image corrupted with a small amount of noise. Even though the filter worked well for uniform regions, its

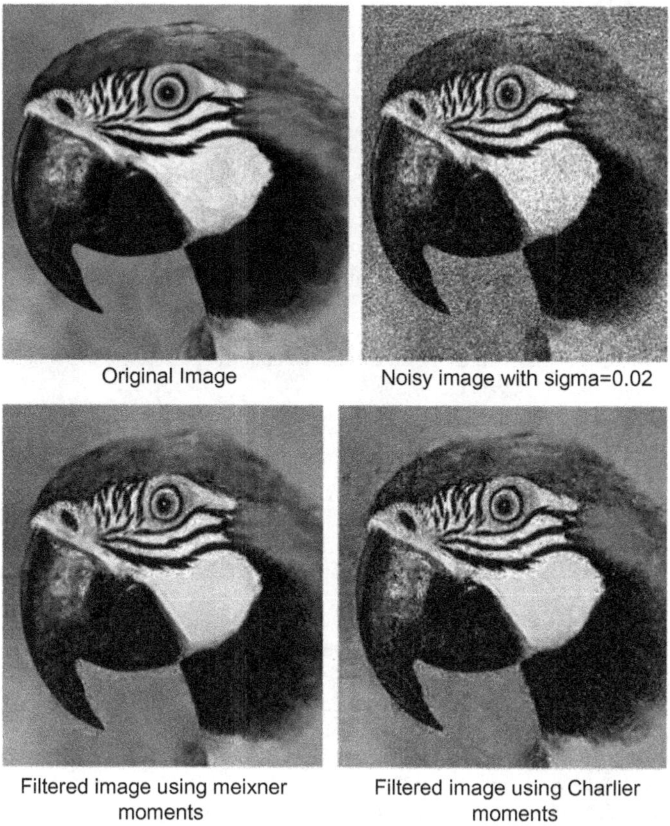

Fig. 7 Filtered of images using Meixner and Charlier moments

performance was not that good at edge locations. This can be solved using a regularization method. Further, the proposed approach worked well for two-class texture segmentation, but its results are not good for four-class problem, this is because the amount of data used for training the network is small. The results are better for K-means and Fuzzy K-means than Kohonen's network. It is observed that there is not much difference between these two moments for selected problems. Finally, further work is required for processing images corrupted with other types of noise like impulse, Poisson, and speckle.

Acknowledgements Author was supported by the University Grants Commission, New Delhi, under Emeritus fellowship. No. F.6-6/2017-18/EMERITUS-2017-18-GEN-10889/(SA-II). Author also thanks Mr. Bharath, Research scholor, ECE, Dept, OU, for assisting in the preparation of the document.

References

1. Zhu, H., Liu, M., Shu, H., Zhang, H., Luo, L.: General form for obtaining discrete orthogonal moments. IET Image Proc. **4**(5), 335–352 (2010)
2. Hmimd, A., Sayyouri, M., Allouchi, H.: Image classification using novel set of Charlier moment invariants. WSEAS Trans. Signal Process. **10**, 156–167 (2014)
3. Hmimid, A., Sayyouri, M., Qjidas, H.: Image classification using separable invariant moments of Charlier-Meixner and support vector machine. Multimed. Tools Appl. (2) (2018)
4. Jahid, T., Hmimid, A., Karmouni, H., Sayyouri, M., Qjidaa, Rezzouk, A.: Image analysis by Meixner moment and a digital filter. Multimedia Tools and Applications (4) (2018)
5. Tuceryan, M.: Moment based texture segmentation. Pattern Recogn. Lett. **15**(7), 659–668 (1994)
6. Victor Marcos, J., Cristobal, G.: Texture classification using discrete Tchebichef moments. J. Opt. Soc. Am. A **30**(8) (2013)
7. Di Ruberto, C., Putzu, L., Rodriguez, G.: Fast and accurate computation of orthogonal moments for texture analysis. Pattern Recogn. **83**, 498–510 (2018)
8. Gelzinis, A., Verikas, A., Bacauskiene, M.: Increasing the discrimination power of the co-occurrence matrix based features. Pattern Recogn. **40**, 2367–2372 (2007)
9. Haralick, R.M., Shanmugam, K., Dinstein, I.: Texture features for image classification. IEEE Trans. Syst. Man Cybern. **3**(6), 610–621 (1973)
10. Liu, L., Chen, J., Fieguth, P., Zhao, G., Chellappa, R., Pietikainen, M.: From Bow to CNN: two decades of texture representation for texture classification. Int. J. Comput. Vis. **127**(1), 74–107 (2019)
11. Ananth Raj, P., Ramana, A.V.: Texture segmentation using orthogonal moments. In: IEEE, INDICON, New Delhi, 15–17 Sept 2006
12. Wang, J., Yin, C.C.: A Zernike moment based non local denoising filter for cryo-EM images. Life Sci. Sci. China **56**(4), 384–390 (2013)
13. Ji, Z., Chen, Q., Sun, Q.S., Xia, D.S.: A moment based non local means algorithm for image denoising. Inf. Process. Lett. **109**, 1238–1244
14. Kumar, Ahlad: Nonlocal means image denoising using orthogonal moments. Appl. Opt. **54**(27), 8154–8160 (2015)
15. Gautschi, W.: On generating orthogonal polynomial. SIAM J. Sci. Stat. Comput. **3**(3) (1982)
16. Sayyouri, M., Hamimid, A., Qjidaa, H.: A fast computation of novel set of Maixner invariant moments for image analysis. Int. J. Ckts, Syst. Signal Process. **34**, 875–900 (2015)
17. Zhang, X., Liao, S.: Color image reconstruction from Charlier moments. J. Theor. Appl. Comput. Sci. **9**(2), 3–13 (2015)
18. Buades, A., Coll, B., Morel, J.M.: A non local algorithm for image de-noising. In: IEEE Conference on Computer Vision and Pattern Recognition, CVPR, vol. 2, pp. 60–65 (2005)
19. https://multibandtexture.recherche.usherbrooke.ca/original_brodatz.html
20. Kylberg, G.: The Kylberg Texture Dataset v. 1.0. Centre for Image Analysis, Swedish University of Agricultural Sciences and Uppsala University, external report No. 35. Available http://www.cb.uu.se/~gustaf/texture/

Tumor Boundary Delineation Using Abnormality Outlining Box Guided Modified GVF Snake Model

Srinivas Thirumala and Srinivasa Rao Chanamallu

Abstract Segmentation is a million dollar task in order to highlight any region of interest (ROI) such as tumor, bleed, edema, or infarct in any medical image. Disease diagnosis, prognosis, surgery, and rehabilitation aspects are guided by segmentation results in oncology. The best ideology is to use high-end active contour models for tumor detection, location, and delineation. Traditional parametric active contour model like GVF (gradient vector flow) snake model suffers from more sensitive parameters, limited capture range, more sensitive to noise, and indentation while segmenting brain tumors in medical images. In order to address and solve these problems, a modified GVF is proposed. The modified snake model proposes a new design for the snake external force to resolve these problems by inducing a strong force near edges. This paper will address how an abnormality outlining box (AOB) guides modified GVF to get rid of demerit like lack of choice of precise initial contour. AOB-guided modified GVF is an integration of AOB and modified GVF external force. The experimental results prove the modified snake model captures the tumor efficiently and quickly even in the presence of any noise artifacts.

Keywords Abnormality outlining box · Brain tumor · Convergence · Deformable models · Delineation · GVF · Modified GVF · Histogram · Medical image segmentation · Parametric active contours · Snakes

1 Evolutionary Past and Introduction

Wide variety of segmentation methods are developed for accurate segmentation of tumors [3] with different imaging modalities. Active contours [1, 4] are suitable for

S. Thirumala (✉)
Department of ECE, Aditya College of Engineering, Surampalem, AP 533437, India
e-mail: tirumala.sri1@gmail.com

S. R. Chanamallu
Department of ECE, University College of Engineering Vizianagaram, JNTUK, Vizianagaram, AP 535003, India
e-mail: ch_rao@rediffmail.com

J. K. Mandal et al. (eds.), *Information, Photonics and Communication*, Lecture Notes in Networks and Systems 79,
https://doi.org/10.1007/978-981-32-9453-0_14

medical image segmentation tasks. In 1988, Terzopoulos et al. introduced deformable models. Deformable model representation is discussed in [4, 7, 9]. Deformable surface is usually described by geometric representation, evolution law, and topology change. Geometric representations may be continuous or discrete. Continuous means either explicit or implicit representation. Implicit surfaces include algebraic and level set whereas explicit surfaces include B-Splines and NURBS. Simple deformable models are point based. Complex models like snakes, GVF, and balloon forces come under parametric whereas level sets are geometric type. Similarly, active shape models (ASM) are based on local (intrinsic) or global (extrinsic) sources/similarity of shape and appearance. Parametric models represent surfaces explicitly in their parametric forms during their deformation with fast convergence. Geometric models handle topological changes naturally. These represent surfaces implicitly as a level set of a higher dimensional scalar function. Original snake model of Kass et al. has limitations like false boundary identification and fails to reach concavities. Appropriate choice of parameters is required for resolving all such problems. Researchers have modified the existing methods or proposed various new models to get rid of the problems of the conventional approach [10–16]. Snake growing method of Berger also introduces similar problems of the original Kass model [3, 17]. An evolved single initial active contour is later translated into multiple snakes by authors of [18, 19]. Next high-energy snakes are ignored and only low-energy snakes are considered [5, 20, 21] as they provide a low energy functional which is required for robust segmentation.

The remainder of the manuscript is divided into four sections. Section 2 gives all the related work and required mathematical background, and Sect. 3 describes the proposed AOB-guided modified GVF implementation details. The results after simulation are postulated in Sect. 4 after tested on different images, and the last Sect. 5 provides conclusion along with a brief view on the future vision of the work.

2 Related Work and Mathematical Foundations

Many external forces like GVF and generalized GVF (GGVF) etc. are available for deformable models. Various snakes such as watershed-based snakes, Kalman snakes, distance snakes, B-spline, G-snakes, and boundary vector field (BVF) snakes are suitable for medical image segmentation solutions. A typical snake is chosen for certain application according to what type of ROI is to be detected and in which modality it is to be operated [22, 23]. It is the basic parametric active contour expressed as $v(s) = (x(s), y(s))$ where $s = [0, 1] \rightarrow R^2$ continuous spatial domain where (x, y) are continuous position variables. The energy functional to be minimized is:

$$E^*_{\text{Snake}} = \int_0^1 E_{\text{snake}}(v(s))\mathrm{d}s = \int_0^1 \left(E_{\text{int}}(v(s)) + E_{\text{img}}(v(s)) + E_{\text{con}}(v(s))\right) \quad (1)$$

Spline internal, image, and external constraint forces energies are denoted by E_{int}, E_{img} and E_{con}, respectively. Internal energy is controlled by tension, rigidity parameter of the contour, i.e., α, β, respectively. External energy is derived from the gray-level image $I(x, y)$, and it pulls the snake toward edges to be highlighted in the image. Next GVF snake is proposed by Xu and prince [2, 7]. Strong forces are induced near edges and by means of a computational diffusion process; edge map extends the gradient into homogeneous regions hence the name GVF. This dense vector external force field is expressed as $V(x, y) = (u(x, y), v(x, y))$ and it minimizes the following energy functional

$$E = \iint \mu\left(u_x^2 + u_y^2 + v_x^2 + v_y^2\right) + f|\nabla f|^2|v - \nabla f|^2 \mathrm{d}x\mathrm{d}y \tag{2}$$

f is edge map and ∇f is its gradient. Euler equations are used for the solution of GVF, viz. GVF Snake which minimizes above energy functional must satisfy (or can be obtained by solving) Euler equation which is a decoupled linear partial differential force balance equation. The weak and strong edges can be discriminated after the computation of the gradient of edge map. A strong edge is one which has a large gradient. The balance between the first and the second terms of above integral is by means of μ which is called as regularization parameter. One can change γ (external force weight) and K(viscosity parameter) along with magnitude of boundary term to achieve appreciable result, but it limits the speed of the algorithm. The mathematical foundations and further information of GVF are given in [7, 10]. Level sets pros. are numerical stability of the solution and easy representation whereas computational complexity is its main con. [18, 19]. The remedy is fast marching and narrow band approach of Adalsteinsson et al. in 1995. The next important model is GGVF Snake. But it can't capture typical irregular boundaries of U- and Ω-shaped objects due to exerted local minimum external forces.

3 Proposed Method

Many of the segmentation methods provide partial segmentation if they are used independently or mutually exclusively. But if they are integrated with modifications, they provide tremendous success just like our modified method. An illustration is our proposed method, viz. AOB-guided Modified GVF.

AOB-guided modified GVF implementation can be broadly divided into two subroutines as shown in above Fig. 1. The first step is to determine the initial contour using AOB implementation, and the next step is to generate modified GVF external force and guide the snake toward tumor boundaries. It can be shown as three aspects with Figs. 2, 3, and 4. First, implement an AOB on test image which covers approximate boundary of the benign or malignant tumors of different size, location, and shape. Malignant tumors are classified as meningioma, glioma, astrocytoma, and metastatic. The structure which is in the shape of a rectangular box is used to find

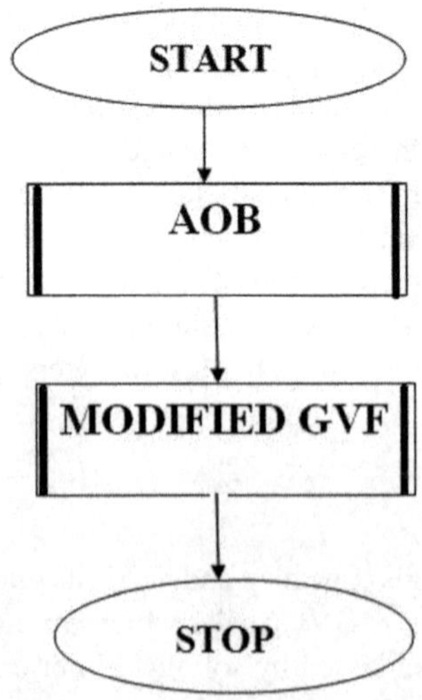

Fig. 1 AOB-guided modified GVF snake model implementation flowchart

the approximate area of abnormality via histograms defined in both directions hence the name AOB. The box located may be a tool for defining initial contour in the forthcoming segmentation step. Actually, it is based on differentiating the left side and right side slices of a brain image as they are dissimilar in case of the presence of any type of abnormality. For this usually, we have considered axial or coronal view brain images. First, divide the image into left and right slices by drawing a partition axis along with highlighting the boundary of the brain image. After this, compare two sections of the pixel on left slice and rotated right slice one by one and vice versa depending on where the abnormality is present in the considered image modality. From the computation of horizontal-oriented and vertical-oriented histograms plots, lower, upper boundaries, and right, left boundaries of abnormality, are obtained respectively [23, 24]. The combination of all the bounds provides the bound box which highlights approximate region or area of abnormality which is termed as Abnormality Approximation Area Aiming (AAAA or A4) Box with two parameters p and q of two histograms for whole scan process of the brain image. These two parameters are the basis for defining the best initial contour in order to obtain the best final contour to fit ROI boundaries, viz. tumor. It is also possible to identify the type of stroke as hemorrhagic or ischemic from the spatial information of histogram within dearth analysis.

Next modified force which diffuses the gradient force toward the object boundary via slight modifications in GVF is used in order to overcome the false contour delineation drawback of GVF. Actually in conventional GVF, the snake is purely closed, but in our novel method, the snake is defined and implemented in such a way that it is not purely closed; i.e., it has redundant cosine and sine terms which lead to create a new external force to detect tumor boundaries accurately as given in Eqs. (3) and (4) in continuous time domain. AOB-guided Modified GVF is represented by following Eqs. (3) and (4).

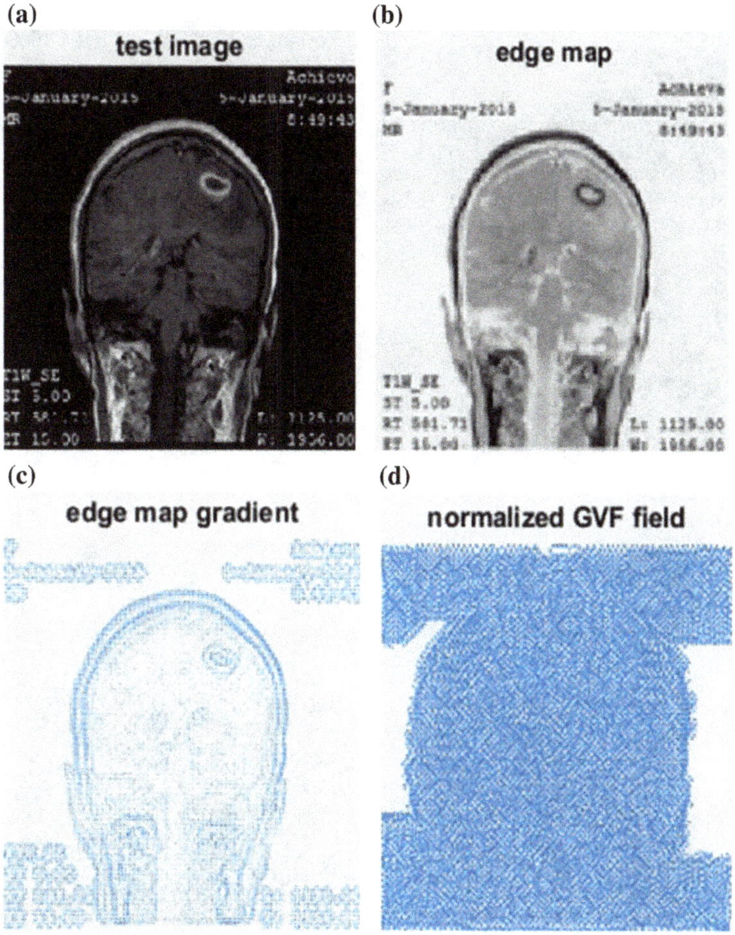

Fig. 2 Steps of the actual GVF process

$$x = a + [k\cos(t) - m\sin(t)] \tag{3}$$

$$y = b + [l\sin(t) - n\cos(t)] \tag{4}$$

Whereas traditional GVF is expressed as

$$(x - a)^2 + (y - b)^2 = k^2 \tag{5}$$

where a, b, k are semi-major and minor axis and radius of evolved contour. k, l, m, and n are strong attraction generation terms to produce deep convolutions. With the computation of the gradient of edge map from Euler equations and by inducing normalized vertical and horizontal modified GVF fields, guides it towards ROI. The

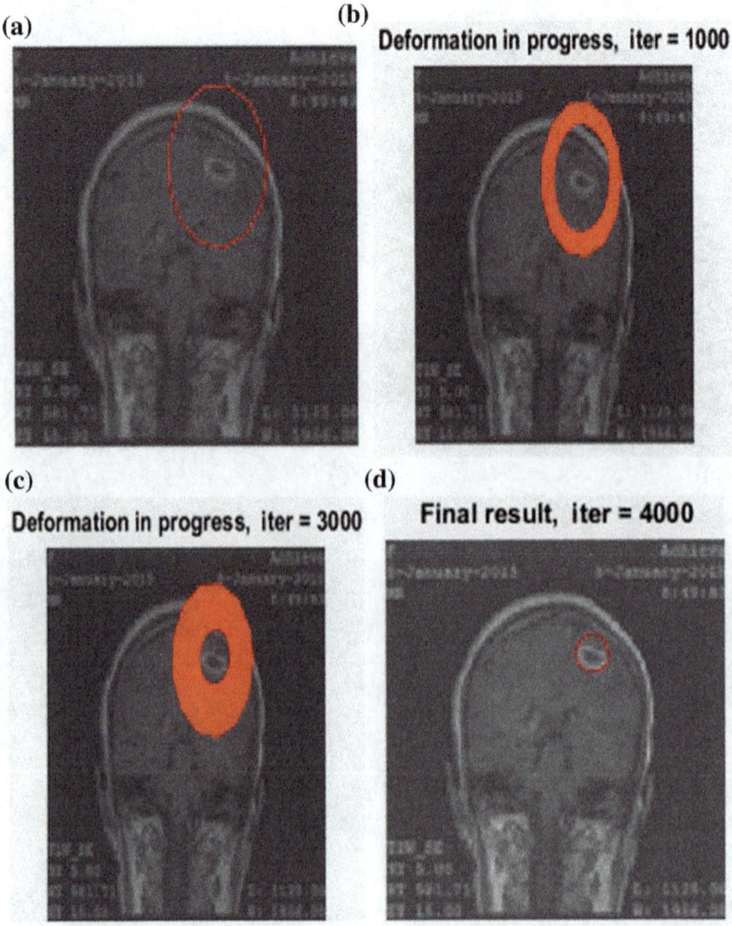

Fig. 3 Brain tumor segmentation using traditional GVF snake model

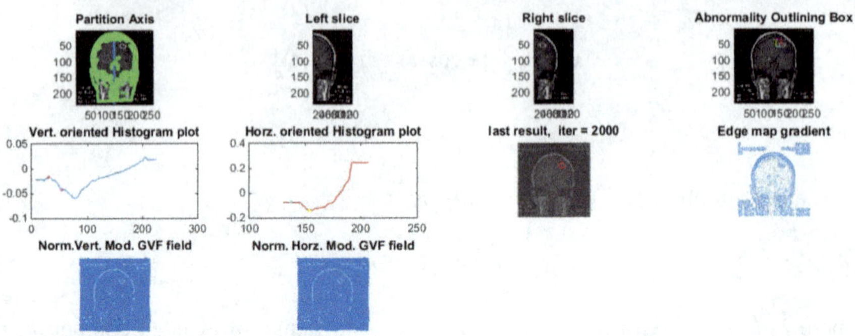

Fig. 4 Steps of AOB-guided modified GVF snake model

time taken for inducing the field twice is close to the time for inducing one field in the GVF. These fields are evolved irrespective of the valley and peak present in the histogram as their corresponding score plots with parameters p, q are used. Next with the contour's modified position along the direction of the defined external force, the new bound term gets changed as given in Eq. (6)

$$\text{new bound term} = \left(u^2 - v^2\right) - 2uv \tag{6}$$

where u, v are vector field parameters of modified GVF. Varying γ, K parameters along with variable magnitude for boundary term causes to achieve appreciable results with high speed using new snake equation of (3), (4) and with Eq. (7).

$$\text{Modmag} = \text{sqrt}\left(u^2 + v^2\right) - (\text{new bound term}) \tag{7}$$

Modified model merits are the easiest way of choosing an initial contour, optimal fit to concave boundaries of the interested lesion, good capture range, and better sensitivity. Hence, this method improves the automatic mechanism for defining initial contour and going toward automatic modeling instead of semiautomatic modeling of the past deformable models; i.e., the level of automaticity is increased in determining the approximate boundaries and later defined snake is deformed to tumor boundary via user inputs same as in semiautomatic modeling.

4 Results and Discussions

This section provides a qualitative and quantitative comparison of AOB-guided modified GVF method versus traditional GVF method. All these experimental results are verified and validated after carrying many experiments based on real medical images using 64-bit windows OS-based Intel i5 processor @ 1.7–2.4 GHz system (with 4 GB RAM) in MATLAB tool. For this, ground truths of GSL Medical College and Hospital—Rajahmundry are considered. Three types of typical brain images with tumors in different locations such as in left bottom direction-oriented, right up direction-oriented, right centered direction-oriented, right bottom direction-oriented tumors are considered.

First results of GVF and later proposed abnormality box guided modified GVF model are shown for magnetic resonance (MR) image with tumor at the right top location. The overall process of traditional GVF snake segmentation can be shown with Figs. 2 and 3. In Fig. 2, GVF external force steps are provided with four subfigures as follows: subfigure a is test image, subfigure b is edge map, subfigure c is gradient of edge map, and final subfigure d is normalized GVF. After this, the intermediate segmentation steps screen shots are shown in Fig. 3 for 1000, 3000, and 4000 iteration deformations using b, c, d subfigures, respectively, and initial contour is shown in subfigure a. The choice of parameter changes initial contour location; hence, many times final segmentation results converge to false boundaries.

Table 1 Comparison of GVF and AOB-guided modified GVF

Image with tumor orientation	Method	No. of iterations	Convergence
Image with right top location-oriented tumor	GVF Snake model	4000	Partial
	AOB-guided Mod. GVF Snake	2000	Excellent

Next, the whole process of AOB-guided Modified GVF can be shown with above direct MATLAB Fig. 4 using ten subfigures referred as a–j in row-wise for discussion. Subfigure a indicates original image with a partition axis along with outlined green outer boundary, subfigures b, c shows left slice with/without rotation, right slice with/without rotation (slice rotation is required in two slices based on the location of tumor in the coronal/axial/sagittal perspective). Subfigure d illustrates AOB after comparing two vertical and horizontal oriented histogram plots (subfigures e and f, respectively). Modified GVF also provides snake equation with a, b, k, l, m, n values for further exact initial contour location. Subfigure g initially shows edge map, and finally, it shows tumor boundary delineation result. Binary edge map is crucial as it is larger in the proximity of edges. Subfigure h indicates edge map gradient. Gradient of edge map is computed which results in efficient anisotropic diffusion process. Subfigures i, j show the normalized modified GVF field in both vertical and horizontal directions, respectively. The results are discussed by a Table 1 with the relative comparison. The first column indicates the location of tumor, the second column represents which method is adopted either traditional GVF or AOB based GVF, the third column tells no of iterations taken for converging toward exact tumor boundary, and the last column indicates degree of convergence as partial/excellent. The choice of parameters considered for experimentation is $K = 0.7$, $\alpha = 3$, $\beta = 0$, $\mu = 2$ and $\gamma = 4$. Appreciable results have been obtained in a short time span. The proposed model only converges for 2000 iterations as in subfigure g of Fig. 4 (Table 1).

5 Conclusion and Future Scope

In our proposed work, conventional GVF is modified to segment brain tumor images. This method even captures minute features exactly from any modality images as compared to ordinary GVF snake method of Xu and Prince. The adoption of the modified vector field equation in conjunction with an adequate selection of regulatory parameters gives us these promising results. Initially, the user can initialize the snake with a desired radius with the guidance of AOB and then selects optimal parameters according to initialization; the snake deforms quickly and very efficiently and gives unbelievable results. These results are used in Macro pathology, and this data is also used by Neurologist, radiotherapist or radiologist for further diagnosis and prognosis. This technique is also used for segmentation of liver tumors of various modalities

[6]. This new modified GVF is combined with another external force called BVF (boundary vector field) and Watershed to recognize the complex shape structure in medical as well as mechanical motion applications to study anatomic structures [8].

Acknowledgements Authors are thankful to the GSL Medical College and General Hospital, Rajahmundry, Andhra Pradesh, India for image resources in order to carry out this work.

References

1. Kass, M., Witkin, A., Terzopoulos, D.: Snakes: An active contour models. Int. J. Comput. Vis. **1**(4), 321–331 (1988)
2. Yu Zhong., Jain, A.K., Dubuisson-Jolly, M.-P.: Object tracking using deformable templates. IEEE Trans. Pattern Anal. Mach. Intell. **22**(5), 544–549 (2000). https://doi.org/10.1109/34.857008
3. Pal, N.R., Pal, S.K.: A review on image segmentation techniques. Pattern Recogn. **26**(9), 1277–1294 (1993)
4. Terzopoulos, D., Fleischer, K.: Deformable models. Vis. Comput. **4**(6), 306–331 (1988)
5. Kichenassamy, S., Kumar, A., Oliver, P., Tannenbaum, A., Yezzi, A.: Gradient vector flows and geometric active contours. In: Fifth International Conference on Computer Vision ICCV 95, pp. 810–815. IEEE Computer Society, Washington, DC, USA (1995)
6. Huang, S., Wang, B., Huang, X.: Using GVF snake to segment liver from CT images. In: 3rd IEEE-EMBS International Summer School and Symposium on Medical Devices and Biosensors, pp. 145–148. MIT, Boston, USA (2006)
7. Terzopoulos, D.: On matching deformable models to images: Direct and Iterative Solutions. In: Topical Meeting on Machine Vision. Technical Digest Series, vol. **12**, no. 9, pp. 164–167. Optical Society of America, Washington, DC, Incline Village, Nevada (1987)
8. Sum, K.W., Cheung, P.Y.S.: Boundary vector field for parametric active contours. Pattern Recogn. **40**(6), 1635–1645 (2007)
9. Williams, D.J., Shah, M.: A fast algorithm for active contours and curvature estimation. CVGIP: Image Understanding **55**(1), 14–26 (1992)
10. Xu, C., Prince, J.L.: Snakes, shapes, and gradient vector flow. IEEE Trans. Image Process. **3**, 359–369 (1998)
11. Jehan-Besson, S., Barlaud, M., Aubert, G.: A 3-step Algorithm using region based active contours for video objects detection. EURASIP J. Appl. Sig. Process. **6**, 572–581 (2002)
12. Peterfreund, N.: The velocity snake: deformable Contour for tracking in spatio-velocity space. Comput. Vis. Image Underst. **73**(3), 346–356 (1999)
13. Yu, Z., Bajaj, C.: Image segmentation using gradient vector diffusion and region merging. In: 16th International Conference on Pattern Recognition, vol. 2, pp. 941–944. IEEE CS Press, Quebec, Canada (2002)
14. Ray, N., Acton, S.T., Atles, T., Delange, E.E., Brookeman, J.R.: Merging parametric active contours with in homogeneous image regions for MRI-Based Lung segmentation. IEEE Trans. Med. Imaging **22**(2), 189–199 (2003)
15. Gunn, S.R., Nixon, M.S.: Robust snake implementation: a dual active contour. IEEE Trans. Pattern Anal. Mach. Intell. **19**(1), 63–67 (1997)
16. Zhu, S.C., Yulle, A.: Region competition: unifying snakes, region growing, and Bayes/MDL for multiband image segmentation. IEEE Trans. Pattern Anal. Mach. Intell. **18**(9), 884–900 (1996)
17. Berger, M.O.: Snake growing. In: Berger, M.O. (ed.) Snake Growing. Computer Vision Proceedings. First European Conference on Computer Vision (ECCV'90). Lectures Notes in Computer Science. Springer, Berlin, pp. 570–572. Antibes, France (1990)

18. Malladi, R., Sethian, J.A., Vemuri, B.C.: Shape modeling with front propagation: a level set approach. IEEE Trans. Pattern Anal. Mach. Intell. **17**(2), 158–175 (1995)
19. Han, X., Xu C., Prince, J.: A topology preserving level set method for geometric deformable models. IEEE Trans. Pattern Anal. Mach. Intell. **25**, 755 (2003)
20. Leung Lam, C., Yin Yuen, S.: A fast active cont. algorithm for object tracking in complex background. In: Mertzios, B.G., Liatsis, P. (eds.) Proceedings of IWISP 1996, pp. 165–168. Elsevier Science B.V., Amsterdam (1996)
21. Chan, T.F., Vese, L.A.: Active contours without edges. IEEE Trans. Image Process. **10**(2), 266–277 (2001)
22. Makhanov, S.S.: Active contours in medical image processing: Theory and Applications. In: 5th International Conference on Knowledge and Smart Technology (KST), pp. xviii–xx. IEEE, Chonburi (2013)
23. Nilanjan, R., Russell, G., Albert, M.: Using Symmetry to Detect Abnormalities in Brain MRI. Comput. Soc. India Commun. **31**(19) (2008)
24. Chawla, M., Sharma, S., Sivaswamy, J., Kishore, L.T.: A method for automatic detection and classification of stroke from brain CT images. In: 31st Annual International Conference of the IEEE Engineering in Medicine and Biology Society, pp. 3581–3584. IEEE, Minneapolis, Minnesota, USA (2009)

Detection and Classification of Cervical Spondylosis Using Image Segmentation Techniques

Aniruddha Paul, Aritra Paul and Pramit Brata Chanda

Abstract In this work, we are trying to implement the segmentation of an X-ray image for cervical spondylosis detection. One of the major concerns of that particular disease is faster detection and identification of diseases in previous stages. According to the opinion of clinical experts, today's X-ray tomography method is the most effective technology in medical science domain for easily diagnosis of particular cervical diseases. Segmentation is a kind of approach that is used to identify the unambiguous region from the particular X-ray image. Today, the diagnosis of cervical spondylosis becomes one of the challenging work. MRI and CT scans used by a doctor for manual inspection is already available. So our proposed method which is automatically detect and analyze the cervical spondylosis using morphological segmentation and edge detection and classification-based approach. The results of this study gaining more than 90% accuracy and sensitivity for identifying and classifying the cervical diseases in X-ray images more accurately. Here, the experimental performance shows better PSNR and MSE values for image quality measurement of the detection of cervical spondylosis more accurately.

Keywords Cervical spondylosis · Edge detection · Image classification · Filtering · Morphological operation · Sensitivity

A. Paul (✉) · A. Paul · P. B. Chanda
Computer Science and Engineering, Kalyani Government Engineering College, Kalyani, West Bengal, India
e-mail: aniruddha.socool@gmail.com

A. Paul
e-mail: sonu.aritra@gmail.com

P. B. Chanda
e-mail: pramitcse@gmail.com

J. K. Mandal et al. (eds.), *Information, Photonics and Communication*, Lecture Notes in Networks and Systems 79,
https://doi.org/10.1007/978-981-32-9453-0_15

1 Introduction

Cervical osteoarthritis in neck also known as cervical spondylosis. The "wear and tear" of the tendon and bones found in our cervical spondylosis. The disease is very much well known to old age people and caused by different factors.

In a large number of ways, the cervical spondylosis can express itself. It can cause some neck and neurological problems of the sphincters and middle or the furthest points if there is spinal cord contribution [1]. Some other factors than maturing can build your of cervical spondylosis. Medical image processing is an upcoming area of research for the advancement of new approaches and software releases for analysis. It plays a vital role in disease diagnosis and enhanced in medical area with patient care and helps medical practitioners to make decision in regard to the type of treatment.

Chitte and Gokhale [1] analyze, "Different Methods for Identification and Classification of Cervical Spondylosis". CNN is a technique, which is based on Long and Short term memory logic has been ended up being a powerful tool planned for the course of action estimate issues with spatial data sources, like pictures or chronicles. Yu and Xiang [2] analyze that, in the Chinese analysis, clinical classification of cervical spondylosis mostly include some approaches using fuzzy calculation. The fuzzy logic is used here for classifying cervical spondylosis as a part of some medical expert systems [1]. Anish Jafrin Thilak et al. [3] "Image Using Pro/E and MATLAB". They take some approach of modeling and analysis of FSU. Chudasama et al. try to implement the segmentation approach using morphological and edge detection method as [4] "Image Segmentation using Morphological Operations". On some images for the segmentation process.

1.1 Background Study

Degenerative cervical spondylosis is the very much familiar, moreover asymptomatic situation, happened due to age and some deteriorating changes in the cervical part of the body. Different clinical diseases like axial neck discomfort and cervical irritation can be classified by the cervical spondylosis. A patient of that particular disease mostly having a mixed combination of axial neck discomfort, cervical irritation, and cervical compression.

1.2 Scope of Work

In general traditional diagnosis method, the symptoms are often seen that above situation for taking final decision and some necessary actions to be taken. Careful clinical examination is required for proper diagnosis of diseases, also a thorough interview of the patient and relatives, X-ray, CT, and so forth. However, X-ray is repeatedly used as a complimentary diagnostic instrument in addition to the clinical discovery in the basic-level hospitals. At this point, the cervical spondylosis has different classes; different kinds of cervical spondylosis need to be cured with different treatments and different kinds of cervical spondylosis have to be distinguished from the nominal disease exact changes in the X-ray images. The X-ray images nominal changes make diagnosis visually a critical job, so it needs experienced personnel to solve. Also with it, the problem is unsolved and the potential of diagnosis of the disease has not been explored in this area. So, our proposed methodology of segmentation has certain practical significance.

1.3 Types of Disease

Different types of stages of diseases regarding cervical spondylosis are:

a. ***Postural Neck Pain***
   It occurs due to postural habits for long period of time. When the head alleged onward in poor posture, the spine of the cervical part must maintain the increasing quantity of weight. As a result, it gives additional pressure on ligaments and muscles and causes the postural neck pain.
b. ***Acute Neck Pain***
   It occurs due to unknown cause in a sudden advancement of the neck realizes extraordinary neck acknowledges uncommon neck torment. There are many causes behind the acute neck pain. One may feel toughness and locked condition on neck.
c. ***Cervical Spondylosis***
   It occurs due to 'wear and tear' of Cervical Vertebrae and the intervertebral disease. In a large number of ways, the cervical spondylosis can express itself. It is mostly common and degrades with oldness.
d. ***Neck Osteitis***
   It is aggravation of bone. In general term, Osteitisa is irritation of bone. This type of neck problem mostly depends on individuals' bone structure or the skeleton structure and some other factors of body.

2 Proposed Methodologies

Here, the technique used different steps as like as:

2.1 Read Color CT Scan Image

Here, some color CT scan images are used for this experimental study and display the image in the very first part of our segmentation process. Basically, these images consist of three color components as Red, Green, and Blue.

2.2 Image Pre-processing

Beginning step is the conversion of the color image into grayscale image because processing time taken by color image is more than grayscale image. The next step is resizing the image to a particular size. So, the requirement of preprocessing is to prepare better image quality more and make results of image. Then change the image with its corresponding grayscale image and show the image in our second step of work.

2.3 Normalization of Image

We normalized the image by using some factors and command in our third step.

2.4 Filtering of Image

Filtering is basically used for removing noise from the image. Different types of filtering techniques are used for reducing the noise for gaining better quality image for working. Then, we filter the image using some filtering method in our fourth step of methodology.

2.5 Edge Detected Image

The use of canny edge detector to show the edge for our image in the fifth step of progress and found that canny edge detector is very useful for this purpose. It is an edge detection operator to detect an extensive range of edges in image. It uses a multi-stage algorithm. It almost detects every edge in image. Canny algorithm can identify edges with noise suppressed at the same time.

2.6 *Median Filtered Image*

Here, median filtering is used for removing the rest of the noise present in images. It is a nonlinear digital filtering method and used to decrease noise at the pre-processing stage for improving the result of processing.

2.7 *Morphological Operations*

Here, some morphological operations are used to obtain the morphological image in this work. Erosion is the best outfit as one of the simple operators in morphological operation. It can cause the elements to become lesser in size. Erosion simply corrodes away the borders of the foreground which outcomes in areas of those pixels lesser in size and holes of those areas turned out into larger [5]. Dilation is the best technique for morphological operation [5]. It is basically used on binary image but can be used on grayscale image. It causes the elements to rise in size. The outcome of this operation will progressively increase the borders of the foreground pixels, thus areas raise in size and holes in that part turned out lesser (Fig. 1).

3 Results and Analysis

Here, the results are shown in given images that the results are showing the intermediate steps of segmentation methodology whatever used for experimental study. The experimental result shows the region of diseases or changes in the disease area of the particular X-ray images (Table 1).

The accuracy is measured with confusion matrix for supervised classification of data. The sensitivity, specificity, accuracy parameter shows better classification accuracy with supervised learning methods with more than 90% of accuracy rates. Here, the accuracy of second image is 100% for achieving better accuracy rates more than other images (Table 2).

The segmented cervical image is required for statistical measurement of different parameters then confusion matrix parameters are used for measuring classification accuracy. Here, the early detection of the disease is done properly. The early detection is very important for clinical experts for taking decision about the causes of disease (Fig. 2).

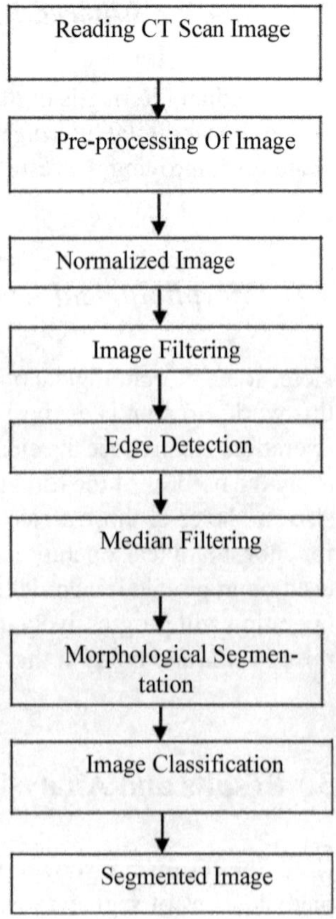

Fig. 1 Proposed segmentation approach

3.1 *Measurement of Image Quality Parameter*

In the experimental phase, we have used the parameter peak signal-to-noise ratio (PSNR) for calculating the difference between the cover image and segmented image. The PSNR for an image of dimension $N \times N$ is given as follows:

$$\text{PSNR} = 10\ \log 10\left(255^2/\text{MSE}\right)(\text{dB})$$
$$\text{where}\quad \text{MSE} = (1/N * N)\sum\sum\left(x_{ij} - x'_{ij}\right)^2. \qquad (1)$$

3.2 *Similarity Index Calculation*

The structural similarity index (SSIM) is a perceptual metric that calculates image quality deprivation produced by handling such as data compression or by sufferers

Table 1 Results of proposed segmentation approach

Images	Input image	Edge detected image	Morphological image	Final segmented image
Cervical spondy-losis image Case 1	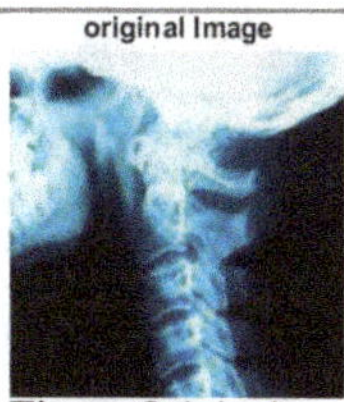 **Fig. a**. Original image [6, 7]	 **Fig. b**. Edge detected image	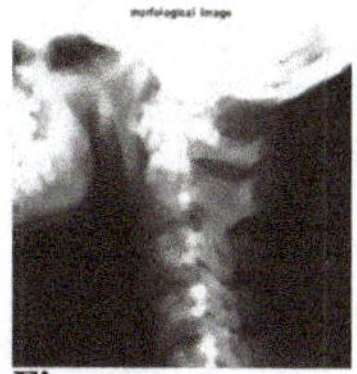 **Fig. c**. Morphological image	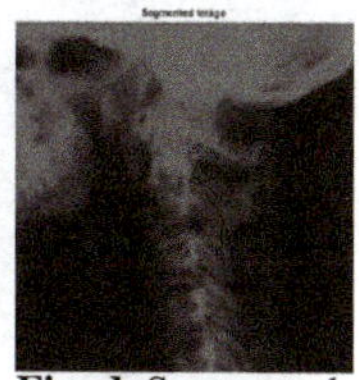 **Fig. d**. Segmented image
Cervical Spondy-losis Image Case 2	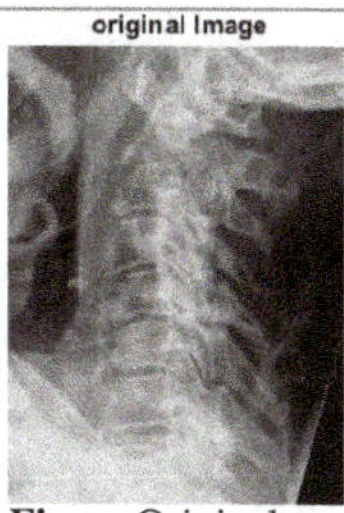 **Fig. a**. Original image[6, 7]	 **Fig. b**. Edge detected image	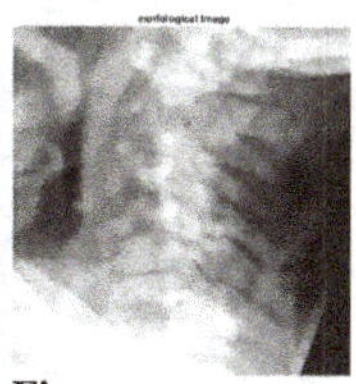 **Fig. c**. Morphological image	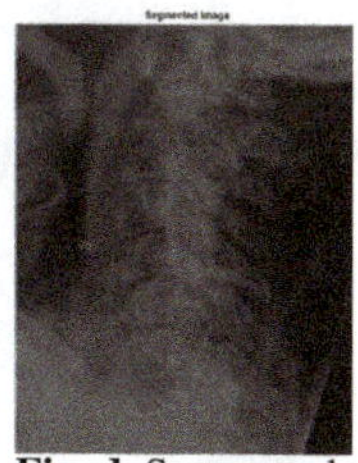 **Fig. d**. Segmented image
Cervical spondy-losis image Case 3	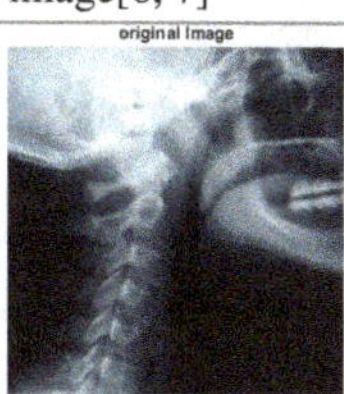 **Fig. a**. Original image[6, 7]	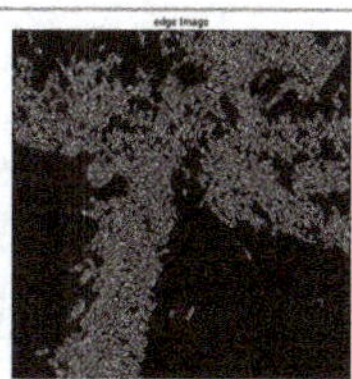 **Fig. b**. Edge detected image	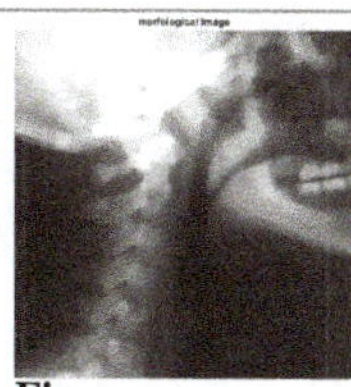 **Fig. c**. Morphological image	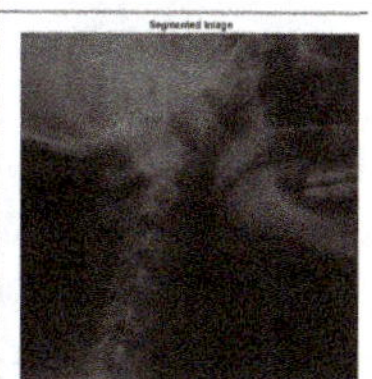 **Fig. d**. Segmented image

Table 2 Several classification parameters in terms of experimental results

Cervical stages of images	False positive rate (%)	Sensitivity (%)	Specificity (%)	Accuracy (%)
Cervical case 1	50.173	99.702	89.835	91.455
Cervical case 2	40.04	100	100	100
Cervical case 3	51.927	99.917	94.807	96.278

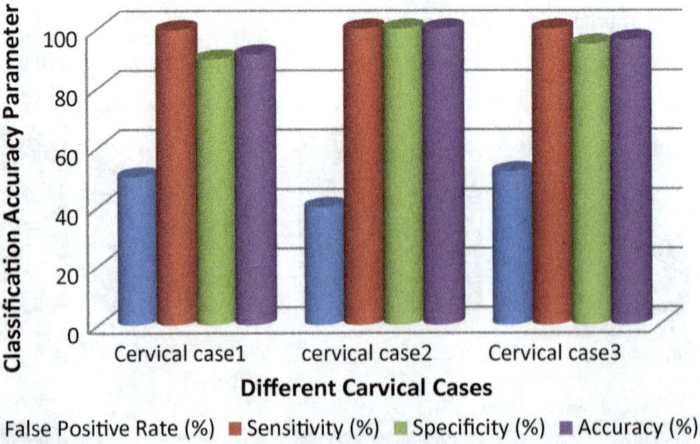

Fig. 2 Classification accuracy parameter

Table 3 Image quality parameters in terms of experimental results

Cervical stages of images	PSNR	MSE	SSIM
Cervical case 1	9.2299	1.124	0.397
Cervical case 2	7.6618	1.114	0.5588
Cervical case 3	9.476	1.336	0.4265

in data conduction. It is a complete reference metric that needs two images from the same image detention—a reference image and a handled image (Table 3).

Our requirement is to get the higher PSNR value and lower MSE value for better quality performance for these cervical images. Here, we got PSNR value of range 7.5–9.5 and MSE value of range 1.1–1.3 for the detection of cervical spondylosis images (Fig. 3).

Here, in the first case, PSNR value is higher and MSE value is lower. In second case, the SSIM value is higher and MSE value is lower, so it will produce the best performance rather than other cases for cervical spondylosis detection.

4 Conclusion

In this particular study, the methods for segmentation of different kinds of cervical spondylosis-disease-related images have discussed. Here the basic concept of Cervical Spondylosis disease, type of neck related diseases, some algorithms related to the problem according to some Literature Review and also Proposed Our Methodology using edge detection and morphological segmentation has been discussed. But from the above review paper, we can conclude that some papers are showing partial automated diagnosis system, some authors tried to increase the efficiency of the images by modifying the segmentation techniques. According to the study of the problem

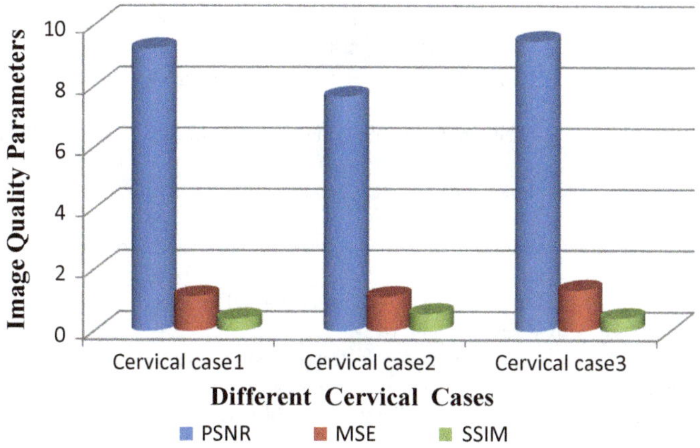

Fig. 3 Image quality parameter

domain, we will try to implement our proposed methodology of cervical spondylosis to the system using edge detection and morphological segmentation that will be more effective and also the accuracy level of the particular set of data will be higher. The used algorithm takes higher than 90% accuracy rates for classification, in terms of better segmentation and classification of cervical diseases. Here, in particular, one case of cervical spondylosis detection is done with higher accuracy and yields better image quality performance than others.

Acknowledgements Authors would like to express our deep and earnest gratitude to Dr. Santanu Banarjee, M.S. (Ortho), Consultant Orthopedic Surgeon for valuable suggestions and support to do this work. We also acknowledge our college, Kalyani government engineering college, and the faculties of computer science and engineering for providing us moral support and the facility of department laboratory. We are also thankful to organizing committee providing the proper support for this purpose. Finally, our thanks go to B. N. Bose Sub Divisional Hospital, Kolkata, for providing me the particular data set of CT scan image which is a very important part of the research study.

References

1. Chitte, P.P., Gokhale, U.M.: Analysis of different methods for identification and classification of cervical spondylosis (CS): a survey. Int. J. Appl. Eng. Res. **12**(21), 11727–11737. ISSN 0973-4562 (2017)
2. Yu, X., Xiang, L.: Classifying cervical spondylosis based on fuzzy calculation, vol. 2014, pp. 1–7 (2014)
3. Anish Jafrin Thilak, J., Suresh, P., Subramani, N., Sathishkumar, S., Arunsankar, V.V.: Analysis of human neck image using Pro/E and MATLAB. Int. J. Innov. Res. Sci. Eng. Technol. **6** (2017). https://doi.org/10.15680/ijirset.2017.0602159
4. Chudasama, D., Patel, T., Joshi, S., Prajapati, G.I.: Image segmentation using morphological operations. Int. J. Comput. Appl. **117**(18), 0975–8887 (2015)

5. Anousouya Devi, M., Ravi, S., Vaishnavi, J., Punitha, S.: Detection of cervical cancer using the image classification algorithms. IJCTA **9**(3) (2016)
6. B. N. Bose Sub Divisional Hospital, Add: B.T. Road, Talpukur, Kolkata, North 24 Parganas, W.B.
7. Dr. Santanu Banarjee, Consultant Orthopedic Surgeon, 54845, Dt. 15-02-99, (W.B.M.C.)
8. Kudva, V., Prasad, K., Guruvare, S.: Detection of specular reflection and segmentation of cervix region in uterine cervix images for cervical cancer screening. IRBM **1**, 3–13 (2017). https://doi.org/10.1016/j.irbm.2017.08.003
9. Jebrietal, B.: Detection of degenerative change in lateral projection cervical spine X-ray images (2015)
10. Sharma, K., Kaur, N.: Comparative analysis of various edge detection techniques. Int. J. Adv. Res. Comput. Sci. Softw. Eng. **3**(12) (2013)
11. Gajpal, T., Meshram, S.: Edge detection technique using hybrid fuzzy logic method. IJERT Int. J. Eng. Res. Technol. **2**(2) (2013)
12. Minavathi, Murali, S., Dinesh, M.S.: Classification of mass in breast ultrasound images using image processing techniques. IJOCA **42**(10) (2012)
13. Kelly, J.C., Groarke, P.J., Butler, J.S., Poynton, A.R., O'Byrne, J.M.: The natural history and clinical syndromes of degenerative cervical spondylosis. Adv. Orthop. **2012** (2012)
14. Bono, C.M., et al.: Diagnosis and treatment of cervical radiculopathy from degenerative disorders. North American Spine Society Evidence Based Clinical Guidelines for Multidisciplinary Spine Care (2010)
15. Rao, R.D., Currier, B.L., Albert, T.J.: Degenerative cervical spondylosis: clinical syndrome, pathogenesis and management. J Bone joint Surg. **89**(6), 1360–1378 (2007)
16. Hochman, M., Tuli, S.: Cervical spondylotic myelopathy: a review. Internet J. Neurol. **4**(1) (2004)

Visual Saliency Based Video Summarization: A Case Study For Preview Video Generation

G. Ramya and Subhash Kulkarni

Abstract A direction research to visual content-driven videos has been in facilitating a short preview of each video through summarization that largely contains short-duration sequence combination of each scene corresponding to stationary camera. This work aims at using visual saliency features to trace scene-change positions in the video. In the present work, visual saliency features are built using color and intensity information as features. Further, using accumulated difference measure (Forward and Backward) in saliency features has been used to filter out false-positive scene-change outcomes. The results have been found to be quite satisfactory and provide closed match to the exact scene-change positions in the video. Significant accuracy is observed with videos using stationary cameras. For moving or non-stationary camera, video summarization has always been a challenging issue. The proposed method has been successfully tested over visual content-driven videos ranging from underwater scenes, fight sequences to surveillance videos in generating preview video.

Keywords Video processing · Visual saliency · Forward and backward accumulated difference (FAD, BAD) · Video summarization · Preview video generation

1 Introduction

Video-based information is seeing an explosive growth on World Wide Web. The video-hosting websites, viz., YouTube, etc., are seeing an upload of more than hundred thousand videos of varying time duration anywhere from one minute to two-plus hours every day. Visual taste has always been subjective among different generations. 95% of the times anyone who watches videos will be resorting to fast-forwarding view. Video can be classified under several categories based on content, activity, emotion, knowledge, etc. Amongst these an interesting class that motivated this present

G. Ramya · S. Kulkarni (✉)
Department of Electronics and Communication Engineering, PESIT-Bangalore South Campus, Bangalore 560100, India
e-mail: sskul@pes.edu

J. K. Mandal et al. (eds.), *Information, Photonics and Communication*, Lecture Notes in Networks and Systems 79,
https://doi.org/10.1007/978-981-32-9453-0_16

work is visual content-driven videos, viz., wildlife videos, physical activity-filled videos, and games' videos like cricket/football, underwater visuals, and surveillance videos.

In [1], authors measured visual saliency by spatiotemporal attention model seeking key frames obtained using geometric features for video summarization. In [2], phase spectrum of Fourier transform (PFT) model is used to calculate the location of salient areas. In [3], audio, visual, and text saliency for video summarization is reported. Based on scene understanding, the authors in [4] proposed a new visual saliency detection algorithm with multiple features for surveillance video. El Khattabi et al. [5] use clustering technique and other techniques for key frame extraction. In [6], key frame selection results into a set of salient frames and video skimming results into a short subclip. Authors in [7] resorted to analyzing and annotating a video sequence of technical talks. A robust motion estimation technique is used to detect key frames. Mu et al. [8] represent video content by a high-dimensional feature curve and detect key frames at curvature points. Ratakonda et al. [9] have presented work based on frame clustering method to select representative frames. In [10] a dynamic neural network is used to select attended locations in order of decreasing saliency. In [11], design and use of steerable filters to compute spatiotemporal orientations are used.

In this proposed work, we present a visual saliency-based approach to detect the key frames and scene-change positions focused on to generate a short preview (video summary) using starting frame of each scene with additional pre-decided number of frames in succession after every key frame and also attentive to detect moving camera scene change with accumulated difference using either forward or backward mode (FAD or BAD) measure on visual saliency features. In this work, the results of video summarization are presented with sample videos taken from YouTube. Comparisons between precise scene change positions of automated results from this work with manual scene change positions are also presented and are quite satisfactory.

The paper is organized by in the following way. Section 2 introduces the saliency feature for video summarization. Section 3 presents the proposed methodology for video summarization. Section 4 validates through results and discussion. Section 5 presents conclusion.

2 Visual Saliency for Summarization

Visual saliency features are obtained by using visual features such as color feature from opponent color space and intensity information.

2.1 Opponent Color Channel Components from Quaternion Representation

RGB image is a true color image where each pixel is a combination of red, green, and blue intensities. *RGB* image is stored in MATLAB as 3D matrix that represents the image height (rows), width (columns) and *r*, *g*, *b* color components for each pixel. Based on opponent color scheme, if *r*, *g*, *b* are red, green, and blue colors, respectively, then opponent color channels are obtained by

$$\begin{aligned} \boldsymbol{RG} &= \boldsymbol{R} - \boldsymbol{G} \\ \boldsymbol{BY} &= \boldsymbol{B} - \boldsymbol{Y} \end{aligned} \tag{1}$$

Where

$$\begin{aligned} R &= r - (g + b)/2 \\ G &= g - (r + b)/2 \\ B &= b - (r + g)/2 \\ Y &= (r + g)/2 - |r - g|/2 - b \end{aligned} \tag{2}$$

2.2 Intensity

Intensity of an image can be obtained by the formula given below

$$I = (r + g + b)/3 \tag{3}$$

2.3 Saliency Features Using Opponent Color Components

In the presence of motion, the color video exhibits a unique opponent color measure in human visual system. This is typically prevalent in red–green, blue–yellow bands, or vice versa. The neuron in visual cortex, if sensitized or excited by red or blue then inhibition is seen in green or yellow band, respectively. This unique characteristic of HVS allows one to compute saliency feature that is unique to each frame. This scalar measure representation against each frame in the video has the ability to characterize motion content which can be compared significantly with conventional spatiotemporal measure [2].

The Quaternion space as in Eq. 2 is an input to opponent color space channel measure as given in Eq. 1. The opponent color channel has the ability to hold static virtual motion information. As motion corresponds to change measure, change detection (scene change) plays a significant role. In order to have increased dynamic range with improved discerning ability of motion, phase spectrum of Fourier transform (PFT) is used. With magnitude replaced by unity, the inverse transform of PFT preserves the motion component significantly. Further energy is distributed in Fourier domain to avoid either loss or isolated singular values using an appropriate low-pass filter, e.g., Gaussian filter. This is further converted into saliency mask using global thresholding. The masked fundamental color channels after vectorization and averaging represent saliency feature vector F_k.

Each frame in the video is replaced by a scalar term called saliency value which is the mean of saliency feature vector given in Eq. 9.

Let RG be the opponent color channel used

$$\begin{aligned} myfft = FT[RG] &= \sum_{\forall(m,n)}\sum RG(m,n)\mathrm{e}^{-j2\pi(mk+nl)} \\ &= |\mathrm{mag}|\mathrm{e}^{j\emptyset} \end{aligned} \tag{2.4}$$

Let $|\mathrm{mag}| = 1$

$$myfft_{ph} = \mathrm{e}^{j\emptyset} \tag{5}$$

$$S_{ph} = \mathrm{IFFT}\left(\mathrm{e}^{j\emptyset}\right) \tag{6}$$

$$S_{\mathrm{mask}} = \mathrm{GlobalThresh}\left(S_{ph} * G_\sigma\right) \tag{7}$$

$$\begin{aligned} F_1 &= S_{\mathrm{mask}} * r \\ F_2 &= S_{\mathrm{mask}} * g \\ F_3 &= S_{\mathrm{mask}} * b \end{aligned} \tag{8}$$

We consider these features as elements of the vector F_K. The saliency value for each frame is created by averaging the computed feature vector.

$$S = \frac{1}{3}\sum_{k=1}^{3} F_k \tag{9}$$

2.4 Saliency Map

Saliency map is an image which represents motion of each pixel distinguishing between stationary and non-stationary objects. Saliency map corresponds to saliency-masked version of r, g, b color channels. It provides a fair indication of motion of

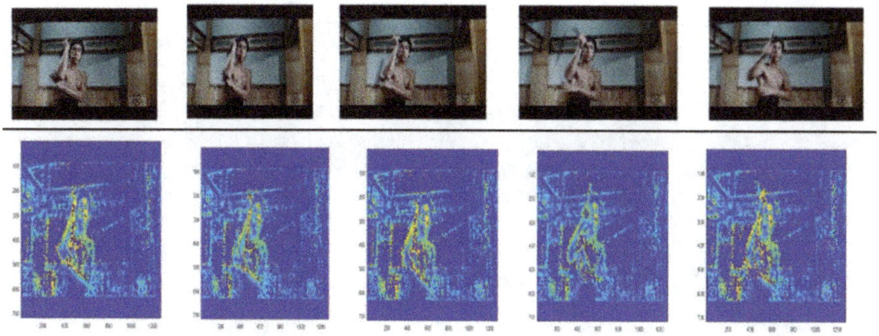

Fig. 1 Original frames of the video clip from movie "Bruce Lee the Fighter" and the corresponding saliency map images

object within the frames. Saliency map is a featured image with color pixels represented as ($F1$, $F2$, $F3$). Few extracted frames from a video and their corresponding saliency maps used in this present work are shown in Fig. 1. Where blue indicates nil or no motion, yellow indicates fair motion.

2.5 *Visual Saliency Curve Representation*

Visual saliency map is a representation of motion of significant pixels in the image. The feature elements in the saliency map also correspond to the opponent color information that sensitize to motion measure via gradient metric. For each frame, saliency features are constructed. Visual saliency value for each frame is obtained by multiplying the scaling factor S with the feature vector, which is a unique single-valued factor that corresponds to the motion dominance within the frame. The visual saliency value for each consecutive frame is plotted as shown in Fig. 2.

2.6 *Mainframes Extraction*

Mainframes are the set of salient images that correspond to significant or peaks in saliency curve plot. In this work, mainframes resemble the attention-seeking signal portions or points of interest. These points are extracted from the visual saliency curve. Scene changes also correspond to sharp transition or maximum change in visual saliency curve, and this can be tracked by analyzing the geometric features of the visual saliency curve. Geometric features like extrema positions (local maxima and minima), peaks and valley, edges and level sets can be used in various video analysis applications.

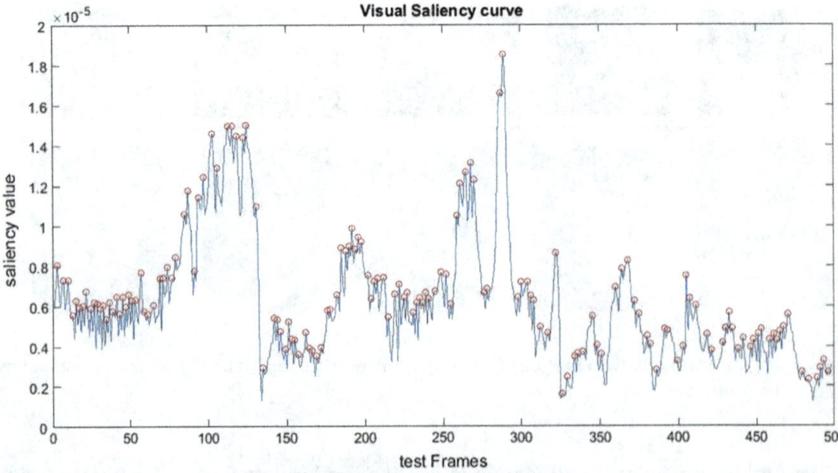

Fig. 2 Visual saliency curve for video "onemin.mp4" having 500 frames. Red circles indicates the peaks of the curve

3 Proposed Methodology for Preview Video Generation

User spends a lot of time in searching the useful videos and for deciding the usefulness one has to watch the entire video. Hence in this present work, we explore on facilitating the user by creating a short preview video for each video through summarization that largely contains short-duration sequence combination of each scene in the video. Based on preview video, users can decide whether to watch full length or to dismiss it by watching the preview video.

For preview video generation, track the scene-change frame positions by thresholding the visual saliency value. We detect the scene change through sharp transitions exhibited in the visual saliency curve. Therefore, we seek those frames which have large transition in saliency value by obtaining the magnitude of difference of visual saliency value of current frame with the saliency value of preceding frame and detect the scene-change frames and sequences through thresholding and removing false positives by using accumulated difference measure of saliency features.

3.1 Algorithmic Steps and Flowchart

STEP 1: Choose the visual content-driven video as an input.
STEP 2: Read the video file to know the properties like duration, frame rate, and total frames present in the video.
STEP 3: Define a variable "total frames" to find the total number of frames present in the input video file.

STEP 4: Compute the visual features of color and intensity using the formulas given in Sect. 2.
STEP 5: Compute the visual saliency map for the frames in the video.
STEP 6: Compute the visual saliency curve by using the color and intensity information and a Gaussian smoother.
STEP 7: Threshold is applied to the difference of consecutive frames of visual saliency curve in order to detect the key frames and these key frames contains frames having higher transitions. Accumulated difference of saliency features is measured to detect the scene-change positions by removing the false positives that are obtained during thresholding.
STEP 8: After detecting the scene change, preview video is generated with a combination of short-duration sequence from each scene present in the video.
STEP 9: Extract the mainframes present in the video from visual saliency curve.
STEP 10: Summarize the video by using mainframes and the resulting summary contains the entire information present in the original video (Fig. 3).

4 Results and Discussion

The videos used in this work are tabulated with name, duration, frame rate, and frames obtained after thresholding in Table 1. Accumulated saliency difference is computed by using seven preceding and seven succeeding frames of each key frame obtained after thresholding. Accumulated difference is average of difference with seven preceding frames or is average of difference with seven succeeding frames that are termed as backward accumulated difference (BAD) and forward accumulated difference (FAD), respectively. In the presence of scene change either in preceding or succeeding frames, along with difference being above a set threshold then such key frames are tagged as possible key frames. When a key frame has no scene change or falling below threshold, then such key frames are summarily rejected as key frame.

Table 2 contains key frames of surveillance video "Road.mp4," and it also contains forward accumulated difference and backward accumulated difference for the given frame number. The video "road.mp4" contains a total of ten scene change at frame numbers 48, 96, 143, 192, 239, 287, 382, 430, 477, and 524 (obtained manually) as shown in Fig. 4 and from the accumulated difference measure, we are able to detect all the scene-change positions at frame numbers 47, 96, 142, 191, 238, 286, 381, 429, 476, and 523 (bold frame numbers in Table 2). Video is summarized by adding consecutive 10–30 frames from each lead scene (key frames) of video "Road.mp4". The original video has 545 frames and preview video is generated using 100 frames with ten consecutive frames chosen for every key frame or 300 frames with 30 frames chosen for every key frame.

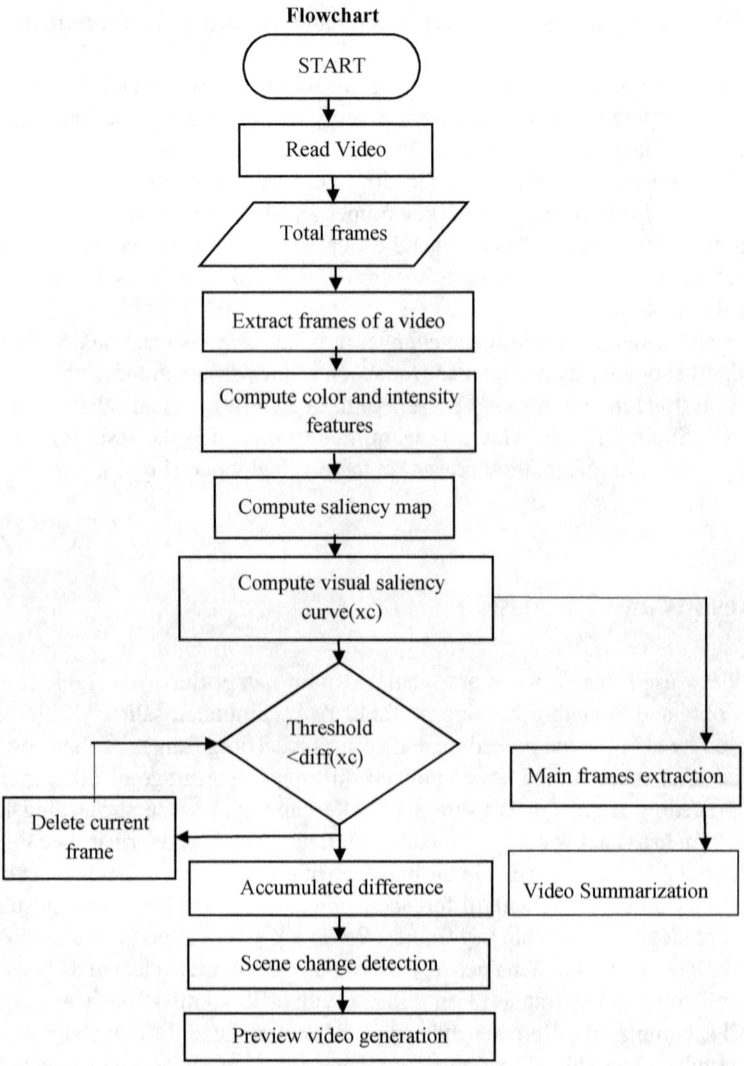

Fig. 3 Algorithmic flowchart

5 Conclusion

In this work, we have presented an efficient algorithm to find the saliency map of each frame and visual saliency curve from the features of visual content-driven videos. We extracted the mainframes from the visual saliency curve through local maxima and local minima. Further, we detected the scene-change position through extracting the key frames of a video by applying threshold to the visual saliency curve and

Table 1 Duration and frame rate of each video used in experimentation and the corresponding key frames obtained after thresholding

File name	Duration (sec)		Frame rate per sec	Total frames	Total no. of scene change	Threshold value	Key frames obtained
	Original size	Summarized preview size (10/key frame)					
Fight.mp4	11	5	23	277	6	$0.2*10^{-4}$	13
Road.mp4	22	8	23	545	10	$0.2*10^{-5}$	19
Cricket.mp4	29	8	25	743	7	$0.25*10^{-5}$	20

Table 2 Forward and backward accumulated difference measure in saliency value for the key frames of "Road.mp4"

File name	Frame no.	BAD	FAD
Road.mp4	**47**	0.3002	**8.0937**
	96	**5.3867**	3.0526
	130	3.9620	4.0375
	142	4.0062	**5.3050**
	167	2.6009	2.7192
	191	2.4369	**4.4955**
	238	0.3474	**4.4768**
	286	0.2494	**3.4294**
	322	0.4221	0.4115
	362	0.3990	0.3792
	381	0.3665	**3.4069**
	429	0.2003	**4.9762**
	446	0.6160	0.5818
	458	0.5720	0.5095
	476	0.4222	**9.1635**
	488	3.4776	3.2309
	504	3.0582	3.5141
	523	3.0341	**6.5455**
	535	0.8648	0.9090

Bold numerals indicate scene change after thresholding (threshold (FAD ~ BAD) > 1.25)

then false positives are removed by using accumulated difference measure and this enables one to construct a video summary (preview video) by reading the starting frames of each scene present in the video, with extension of few additional frames in succession after every key frame.

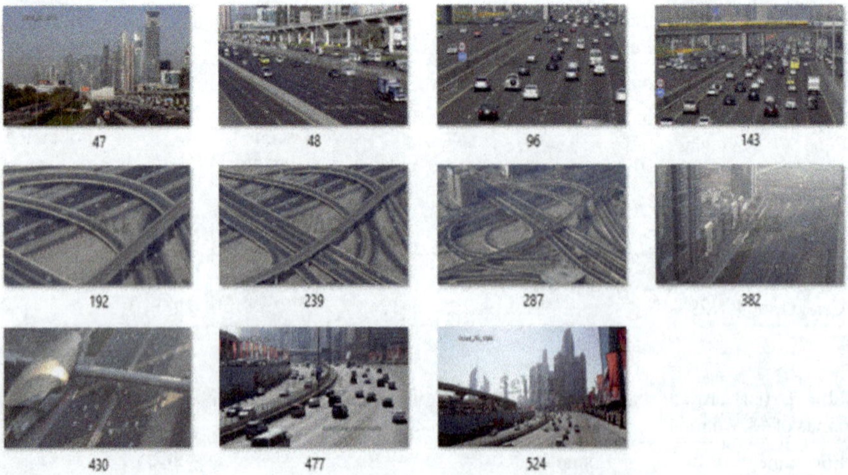

Fig. 4 Actual key frames corresponding to scene change of “Road.mp4”

Future work involves exploring alternate measure other than accumulated difference to identify key frames with increased precision. Another direction could be the smart preview video generation for multimedia content-driven videos.

References

1. Evangelopoulos, G., Rapantzikos, K., Potamianos, A., Maragos, P., Zlatintsi, A., Avrithis, Y.: Audiovisual attention modeling and salient event detection. In: Multimodal Processing and Interaction: Audio, Video, Text (eds.). Springer (2008)
2. Guo, C., Zhang, L.: A novel multiresolution spatiotemporal saliency detection model and its applications in image and video compression. IEEE Trans. Image Proc. **19**(1), (2010)
3. Evangelopoulos, V., Zlatintsi, A., Skoumas, G., Rapantzikos, K., Potamianos, A., Maragos, P., Avrithis, Y.: Video event detection and summarization using audio, visual, text saliency. In IEEE Trans., ICASSP (2009)
4. Tong, Y., Konik, H., Cheikh, F.A., Guraya, F.F.E., Tremeau, A.: Multi-feature based visual saliency detection in surveillance video. In: Visual Communications and Image Processing 2010, vol. 7744 (2010)
5. El Khattabi, Z., Tabii, Y., Benkaddour, A.: Video summarization: techniques and applications. Int. J. Comput. Inf. Eng. **9**(4) (2015)
6. Ying, L., Lee, S.-H., Yeh, C.-H., Kuo, C.-C.: Techniques for movie content analysis and skimming. In IEEE Signal Proc. Mag. **23**(2) (2006)
7. Avrithis, Y., Doulamis, A., Doulamis, N., Kollias, S.: Summarization of video taped presentations: automatic analysis of motion and gesture. Comput. Vision Image Underst **75**(12), 3–24 (1998)
8. Mu, Y., Lu, L., Zhang, H., Li, M.: A user attention model for video summarization. In: Proceedings ACM Int’l Conference on Multimedia (2002)
9. Ratakonda, K., Sezan, M., Crinon, R.: Hierarchical video summarization. In Proceedings SPIE, Visual Communication and Image Processing, vol. 3653, pp. 1531–1541, (Dec 1998)

10. Itti, L., Koch, C., Niebur, E.: A model of saliency based visual attention for rapid scene analysis. IEEE Trans. PAMI **20**(11), 1254–1259 (1998)
11. Freeman, W.T., Adelson, E.H.: The design and use of steerable filters. IEEE Trans. PAMI **9**, 891–906 (1991)

A Moment Based Feature Extraction for Texture Image Retrieval

Ivy Majumdar, B. N. Chatterji and Avijit Kar

Abstract The most important tasks for texture-based image retrieval system are efficient method of feature extraction. Invariant image moments, viz., geometric moment (GM) and Zernike moment (ZM) are used for the feature extraction in content-based image retrieval (CBIR) system. GM is not orthogonal, and ZM is orthogonal, but it shows poor reconstruction of images. Hence, they are not fit enough for texture-feature-based image retrieval system. Hence, an efficient texture feature extraction algorithm having low computational time and high retrieval accuracy is required. Legendre moment (LM) is an orthogonal moment and represents any image with almost nil information redundancy. Different methods for the computation of Legendre moment are available. In this paper, a fast and accurate method of calculating LM called exact Legendre moment (ELM) proposed by Hosny is used for feature extraction. Comparison has been carried out among the method described in this paper, ZM and GM. Performance of the proposed method in the presence of Gaussian noise also has been tested and compared with other moment-based feature extraction systems.

Keywords Texture-based image retrieval · Feature extraction · Zernike moment · Exact Legendre moment

1 Introduction

On the basis of periodicity, texture is broadly categorized into two types: homogeneous and inhomogeneous. Homogeneous texture can be represented using various descriptors, e.g., co-occurrence matrix, histogram, etc. But, they require a complex computation. A comprehensive literature review has been provided by Majumdar et al. [1] regarding texture feature extraction.

I. Majumdar (✉) · B. N. Chatterji
B.P. Poddar Institute of Management and Technology, Kolkata, West Bengal, India
e-mail: ivy_majumdar@yahoo.co.in

A. Kar
Jadavpur University, Kolkata, West Bengal, India

J. K. Mandal et al. (eds.), *Information, Photonics and Communication*,
Lecture Notes in Networks and Systems 79,
https://doi.org/10.1007/978-981-32-9453-0_17

After introduction of invariant by Hu [2], moments they are widely used in image analysis. Mihran [3] used geometric moment (GM) for texture features extraction directly from the gray-level image in local regions. But, it is not orthogonal and hence gives poor representation of the images. Orthogonal moments like Zernike moment (ZM) have been used by Hitam et al. [4] for CBIR. ZM shows good performance but has high computational complexity. Major limitation of ZM is that it is calculated with certain approximation considering a unit circle [5]. Due to this approximation, retrieval accuracy is very low. Hence in this paper, another orthogonal moment, viz., LM has been considered to represent the image effectively.

LM can be computed using different algorithms. Mainly two-dimensional geometric moments are dealt with these methods. During calculation of LM few approximations are adopted, hence the result may be erroneous. Situation becomes critical with an increase in the order of the moment. These algorithms are time-consuming also. A fast and perfect method was projected by Hosny [6] for LM computation known as exact Legendre moments (ELM). The effectiveness of ELM with the strength of fuzzy relationship (SFR) for texture feature extraction has been observed in this paper.

Efficiency of the proposed ELM-based feature extraction system for noisy images has also been explored in this.

2 Overview on Different Moments

Feature extraction is defined as a process that converts images into unique, distinctive, and compact form. According to studies, a set of moments M_{pq} can be uniquely determined by $f(x, y)$. Orthogonality of moments leads to low information redundancy and hence improved computation efficiency. Moments are widely used in various fields of image processing, viz., pattern recognition, face recognition, image reconstruction, image classification and indexing, and target identification and image analysis.

2.1 *Geometric Moment (GM)*

The two-dimensional geometric moment (GM) of image intensity function $f(x, y)$ for the order $(p + q)$ is given as

$$M_{pq} = \int_{-1}^{1} \int_{-1}^{1} x^p y^q f(x, y) \mathrm{d}x \mathrm{d}y \tag{1}$$

Here $p, q = 0, 1, 2, \ldots$ Image function $f(x, y)$ is expressed on the square with dimension $[-1, 1] \times [-1, 1]$. For a digitized image $f(x_i, y_j)$ of dimension $M \times N$, Eq. (1) is approximated by the following equation:

$$\widetilde{M}_{pq} = \sum_{i=1}^{M} \sum_{j=1}^{N} x_i^p y_j^q f(x_i, y_j) \Delta x \Delta y \tag{2}$$

where $x = x_i - x_{i-1}$, $y = y_j - y_{j-1}$ are taken as sampling intervals along x and y directions. The moment in Eq. (2) is not rotation, translation, and scale-invariant.

Expression given by the Eq. (3) can be used to achieve translation invariance central moment

$$\mu_{pq} = \sum \sum (x - \bar{x})^p (y - \bar{y})^q f(x_i, y_j) \tag{3}$$

where $\bar{x} = \frac{m_{10}}{m_{00}}$ and $\bar{y} = \frac{m_{01}}{m_{00}}$.

2.2 Zernike Moment (ZM)

In 1980, Teague [7] proposed Zernike polynomial-based system for image processing. Zernike [8] introduced a complete set of simple rotation properties' based orthogonal functions over the interior of the unit circle ($x^2 + y^2 = 1$). The polynomial is given by the following equation.

$$V_{nm}(x, y) = V_{nm}(\rho \sin \theta,\ \rho \cos \theta) = R_{nm}(\rho) e^{jm\theta} \tag{4}$$

where n may be zero or any positive integer, m may be positive or negative integers depending on $n - |m|$ is even or $|m| \leq n$, θ being the angle of the vector with x-axis in counterclockwise direction, ρ is the distance between the origin and the pixel (x, y) (length of the vector).

The radial polynomial $R_{nm}(\rho)$ is given as

$$R_{nm}(\rho) = \sum_{s=0}^{(n-|m|)/2} (-1)^s \frac{(n-s)!}{s! \left(\frac{n+|m|}{2} - s \right)! \left(\frac{n-|m|}{2} - s \right)!} \rho^{n-2s} \tag{5}$$

2.3 Legendre Moment (LM)

2D Legendre moment with intensity value $f(x, y)$ for the order $p + q$ is:

$$L_{pq} = \frac{(2p+1)}{2} \frac{(2q+1)}{2} \int_{-1}^{1} \int_{-1}^{1} P_p(x) P_q(y) f(x, y) \mathrm{d}x \mathrm{d}y \tag{6}$$

Kernel function $P_p(x)$ depicts pth order LP. It is expressed as

$$P_p(x) = \sum_{k=0}^{p} a_{kp}x^k = \frac{1}{2^p p!}\left(\frac{\mathrm{d}}{\mathrm{d}x}\right)^p\left[\left(x^2 - 1\right)^p\right] \quad (7)$$

where $x \in [-1, 1]$ and $P_p(x)$ follows the recursive function given below.

$$P_{p+1}(x) = \frac{2p+1}{p+1}xP_p(x) - \frac{p}{p+1}P_{p-1}(x) \quad (8)$$

With $P_0(x) = 1, P_1(x) = x$ and $P > 1$.

Hosny [6] projected a perfect and fast process to estimate exact Legendre moment for $M \times N$ image as follows:

Rewriting Eq. (9), we get:

$$\tilde{L}_{pq} = \sum_{i=1}^{n}\sum_{j=1}^{n} I_p(x_i)I_q(y_j)f(x_i, y_j) \quad (9)$$

$$\hat{L}_{pq} = \sum_{i=1}^{M} I_p(x_1)Y_{iq} \quad (10)$$

Y_{iq} represents the qth order moment of ith row.

3 Exact Legendre Moment-Based System

For real-time application, texture image retrieval system must be very fast. Daubechies orthogonal wavelet [9] with pyramidal wavelet transform (PWT) has been applied to compress the images before feature extraction using ELM. Figure 1 illustrates the basic architecture of the proposed system.

The proposed method follows the steps described below:

Step 1. PWT is applied to all of the database images up to third level.
Step 2. ELM is calculated for the compressed images and feature database is created.
Step 3. Feature vector is extracted for the query image using ELM following Steps 1 and 2.
Step 4. Strength of fuzzy relation between query image and database images are computed as given equation.

$$r_{ij} = \frac{\left(\mathrm{Min}\left(\mu_{ij}(A), \mu_{ij}(B)\right)\right)}{\left(\mathrm{Max}\left(\mu_{ij}(A), \mu_{ij}(B)\right)\right)} \quad (11)$$

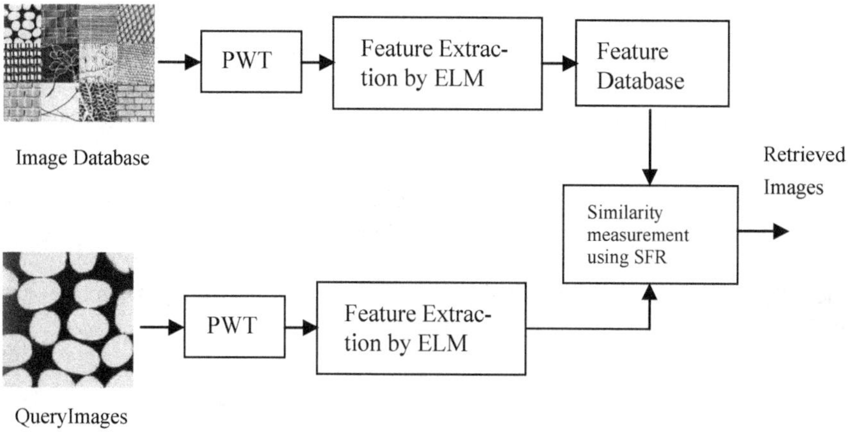

Fig. 1 Proposed image retrieval system using ELM

r_{ij} is the strength of relationship between A_{ij} and B_{ij}. The strength of relationship between A and B (S_{AB}) can be obtained as follows

$$S_{AB} = \sum_{i=1}^{N} \sum_{j=1}^{N} \frac{r_{ij}}{N^2} \tag{12}$$

Step 5. Images are ranked according to the SFR and retrieved.

4 Experimental Result

A total of 111 textures images of size 540 × 540 from Brodatz texture album [10] have been considered for database creation. Each image is equally divided into sixteen 160 × 160 non-overlapping subimages given in Fig. 2 providing a database of total 1776 images.

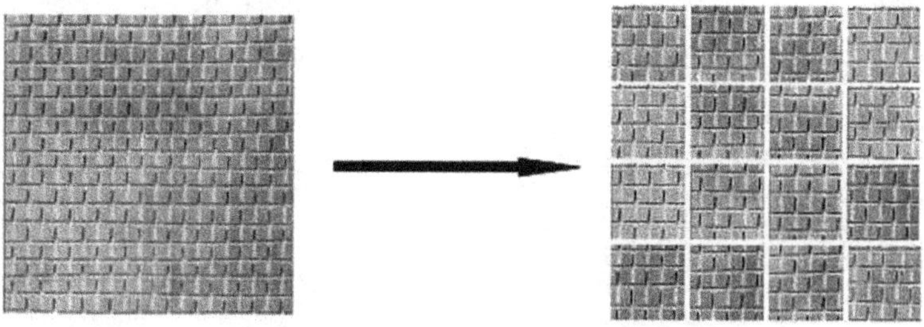

Fig. 2 Example of images (D1) from Brodatz album divided into 16 equal subimages for database creation

Each query image has 16 similar images which are the part of the same image depicted as S_q. Performance has been measured by average retrieval accuracy.

This is the number of ground truth images retrieved in the first 16 images. So, retrieval accuracy RA_q is computed as:

$$RA_q = \frac{\| S'_q \|}{\| S_q \|} \tag{13}$$

where $S'_q \subset S_q$. Overall average retrieval accuracy for N query images can be calculated as

$$\text{Average Retrieval Accuracy} = \frac{1}{N}\sum_{q=1}^{N} RA_q \tag{14}$$

Moments of the order 8 starting from 4 are considered. Dimensions of corresponding feature vectors are 15, 21, 28, 36, and 45 for ELM (the feature vector length up to 8th order for ZM and ELM are shown in Table 1). GM has feature vector of length 7 for all the order.

The time of computation increases with the length of the feature vector. To reduce the computation time, we have used pyramidal wavelet transform (PWT) up to third level, which compressed the size of the original image by one sixty-fourth, and then calculate the moments of the compressed images. Reduction in the size of the images causes much reduced computation time.

Table 2 compares the average retrieval accuracy of the projected texture image retrieval system and other GM- and ZM-based texture image retrieval systems. The table depicted that the system with eighth-order ELM gives the best result with

Table 1 Order of Zernike moment and Legendre moments up to order 8

Order	No. of moments in each order	Total no. of moments up to order $(p + q)$ for Zernike moment	Total no. of moments up to order $(p + q)$ for Legendre moment
0	1	1	1
1	1	2	3
2	2	4	6
3	2	6	10
4	3	9	15
5	3	12	21
6	4	16	28
7	4	20	36
8	5	25	45

Table 2 Average retrieval accuracy for different moments

Methods	Moments order				
	4	5	6	7	8
Geometric moment	56.81	57.81	58.72	59.15	59.56
Zernike moment	62.15	62.81	64.56	65.15	66.81
exact Legendre moment	71.56	71.81	72.81	73.72	**74.15**

retrieval efficiency 74.15% followed by ZM (66.81%) and GM (59.56%). For texture image retrieval system with fourth order, the retrieval efficiency for ELM, ZM, and GM are 71.56, 62.15, and 56.81%, respectively. So from the table, we can conclude that for any order of the moments, ELM-based texture feature gives the best retrieval performance (marked bold in the Table 2).

Figure 3 illustrates an average retrieval rate for different numbers of top retrieved images with different moments with order number 8. It is clear from the figure that ELM gives retrieval efficiency 77.47% for first 20 retrieved images and 88.67% for first 50 retrieved images. For ZM, these values are 69.66 and 76.56% and for GM, it is 61.16 and 63.95%, respectively. We can infer from these data of Fig. 3 that the retrieval efficiency of all moments increases with the increasing number of retrieved images with ELM the best performance.

The experiments have been carried out with MATLAB7.1 on Intel Core2 Duo CPU E7500, 2.93 GHz, 2 GB RAM, Microsoft Windows XP Professional Service Pack 2 machine.

Table 3 provides feature length and CPU times for feature search of the database for a given query image. GM is the fastest method, but its retrieval performance is

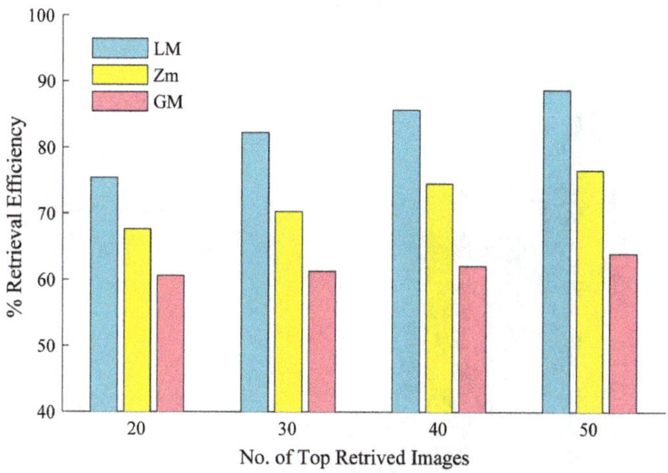

Fig. 3 Average retrieval accuracy with different numbers of retrieved images

Table 3 Time taken for feature extraction and searching

Process	GM		ZM		LM	
Order	4th	8th	4th	8th	4th	8th
Length of feature vector	7	7	9	25	15	45
Searching and retrieval (s)	0.49	0.49	0.95	1.25	0.536	0.849

not good. Time taken by ZM is maximum due to its computational complexity. ELM is comparatively fast as well as gives the best retrieval performance.

Performance of the proposed method has been evaluated under noisy environment. In this case, dataset of 720 images from Brodtz Album [11] was used. White Gaussian noise was introduced to the query images. Then, feature vectors were created following the steps discussed in Sect. 3. The tests were conducted by increasing introduced noise in two ways. In the first case Gaussian noise with zero mean (variance up to 0.2) white noise was introduced to the query images.

Figure 4 demonstrates the average retrieval accuracy under first noisy environment with different orders of ELM (considering top 50 image retrieval). Eighth order shows about 50% retrieval accuracy up to 0.2 variance, whereas fourth order falls below 50% at about 0.06 variance.

In the second case, both Gaussian mean and variance were kept changing. It is clear from Fig. 5 that up to mean 0.02 and variance 0.03, the system gives 50% average retrieval accuracy for fourth-order ELM. Above that noise level, the retrieval accuracy for eighth-order ELM is more or less 50%, but for other order, it falls below 50%.

Figure 6 shows the first 20 images those are retrieved for noisy image D1 with white Gaussian noise (having 0.3 mean and 0.4 variance). We can see from the figure that even for noisy image ELM gives better performance than the ZM.

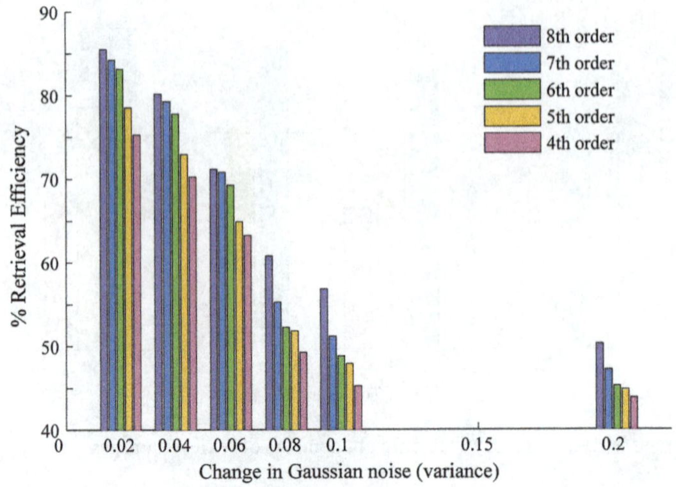

Fig. 4 Average retrieval accuracy for zero mean different variance Gaussian noise with different orders

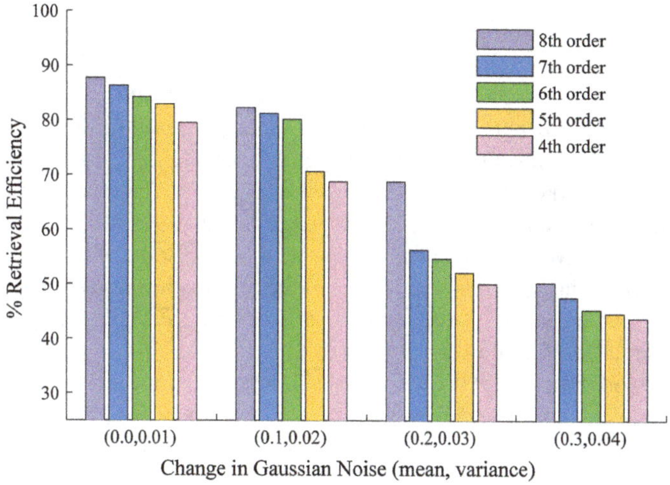

Fig. 5 Average retrieval accuracy with Gaussian noise for different means and variances

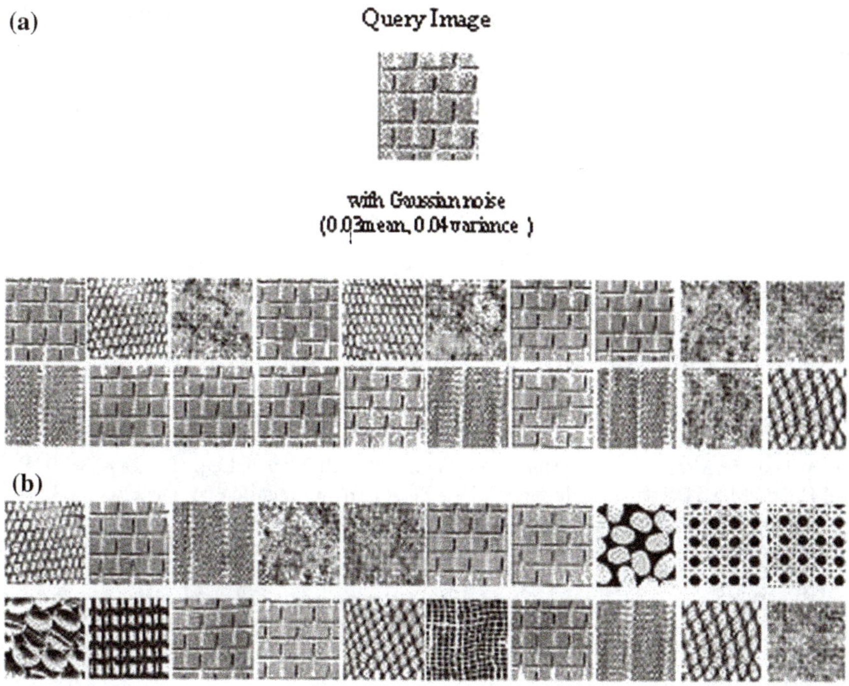

Fig. 6 Retrieval example for noisy query image D1 using **a** exact Legendre moment **b** Zernike moment

5 Conclusion

High retrieval accuracy and less computing time is the requirement for real-time CBIR system. In this paper, a novel system has been proposed for feature extraction and texture image retrieval using DWT and ELM. A large texture database of 1776 images is used to carry out the experiments in our system. The result shows that the average retrieval accuracy of ELM-based texture image retrieval system is superior to both GM- and ZM-based systems. In our method, we have used standard DWT up to third level. It has compressed the size of the original image by one sixty-fourth and hence has reduced the feature extraction time using ELM in comparison with that of the direct ELM calculation time.

Previous work reported by Kokre et al. [11] indicates that combination of energy and standard deviation gives the best retrieval performance (71.71%) than other feature vectors. From our experimental results, we can conclude that the proposed method using ELM as feature vector provides better retrieval accuracy (71.81% with fifth order onwards).

To estimate the performance of the proposed system with ELM as feature vector under noisy environment experiments with noisy query images (having Gaussian noise introduced in them) also have been conducted. For experiments with noise, we have used database D1 consisting of 720 images. The result shows that the noisy texture images even can be identified and matched and can be retrieved efficiently up to a certain noise level, which is also a requirement for a good texture image retrieval process.

References

1. Majumdar, I., Kar, A., Chatterji, B.N.: Texture feature matching methods for content based image retrieval. IETE Tech. Rev. **24**(4), 257–269 (2007)
2. Hosny, K.M.: Exact Legendre moment computation for gray level image. Pattern Recogn. **40**, 3597–3605 (2007)
3. Mihran, T.: Moment based texture segmentation. Pattern Recogn. Lett. **15**, 659–668 (1994)
4. Hitam, M.S., Bakar, S.A., Jawahir, W.N., Yussof, W.: Content based image retrieval using Zernike moments for binary and grayscale images. In: Papakostas, G.A. (ed.) Moments and Moment Invariants—Theory and Applications, vol. 1, chap. 12. GCSR Vol. 1, © Science Gate Publishing (2014)
5. Kim, W.Y., Kim, Y.S.: A region based shape descriptor using Zernike moments. Sig. Process. Image Commun. **16**, 95–102 (2000)
6. Hosny, K.M.: Exact Legendre moment computation for gray level images. Pattern Recogn. **40**, 3597–3605 (2007)
7. Teague, M.R.: Image analysis via the general theory of moments. J. Opt. Soc. Am **70**(8), 920–930 (1980)
8. Zernike, F.: Beugungstheorie des schneidenver-fahrens und seiner verbesserten form, der phasenkontrastmethode. Physica **1**, 689 (1934)

9. Daubechies, I.: The wavelet transform, time-frequency localisation and signal analysis. IEEE Trans. Inf. Theory **36**, 961–1005 (1990)
10. Brodatz, P.: Textures: A Photographic Album for Artists and Designers. Dover, New York (1966)
11. Kokare, M., Chatterji, B.N., Biswas, P.K.: Wavelet transform based texture features for content-based image retrieval. In: Proceedings of 9th National Conference on Communications, pp. 443–447 (2003)

A Novel Approach to Compression of Satellite Images Using Butterworth Filtering

Anirban Patra, Swagata Bandyopadhyay, Debasish Chakraborty and Arijit Saha

Abstract Compression of images is an important application in the field of satellite image processing as it is suitable for optimization of storage space and sharing over the Internet with optimum bandwidth. In our present communication, a new method for compression of satellite images using Butterworth low-pass filtering has been proposed. Initially, we have transformed the selected satellite image into frequency domain. By varying the cutoff frequency, satellite image has been filtered out. We have used three different cutoff frequencies to extract image from its spectrum. However, to get a reasonably good output, information loss of images should be minimized during compression. To measure the quality of the output images, PSNR values have been calculated in each case.

Keywords Image compression · Fourier domain · Butterworth filtering · LISS III 23 m sensor

1 Introduction

Different techniques are used for compression of satellite images either directly from the image or from transformed part of the images. For compression of satellite images, integer form of wavelet regression was described by Md. Al Mamun and Md. Ali Hossain. By modifying the value of temporal correlation, they have achieved better compression ratio [1]. Compression of satellite images is also possible by removing noise bit from images. This method was clearly described by K. Sahnoun and N.

A. Patra (✉) · S. Bandyopadhyay
Department of ECE, JIS College of Engineering, Kalyani, West Bengal, India
e-mail: anitublu@gmail.com

D. Chakraborty
RRSC-East, ISRO, Kolkata, West Bengal, India
e-mail: deba.isro@gmail.com

A. Saha
Department of ECE, BPPIMT, Kolkata, West Bengal, India
e-mail: arijit_sh@yahoo.com

J. K. Mandal et al. (eds.), *Information, Photonics and Communication*,
Lecture Notes in Networks and Systems 79,
https://doi.org/10.1007/978-981-32-9453-0_18

Benabadji in their research paper. In their work, they have used discrete wavelet transform for noise removal to compress satellite images [2]. K. Sahnoun and N. Benabadji proposed a method of image compression based on evidence theory. They have used k-nearest neighbor (kNN) algorithm for compression [3]. They have also used a different technique for compression, based on Fourier transform and Huffman coding [4]. T. Memane and S. D. Ruikar used discrete wavelet transform in their work. They have analyzed the performance of different wavelets for satellite image compression [5]. R. M. Susilo et al. described a method of image compression of X-sat images. In their system, they have used only hardware-based solutions in this lossless compression technique [6]. I. Hacihaliloglu and M. Kartal have used discrete cosine transform for lossless image compression [7].

In our present paper, we have applied Butterworth low-pass filter in Fourier domain for image compression. To do this, we have taken the Fourier spectrum of the image. This is done to get detail information. In the spectrum plane, frequency in the center region is zero (i.e., dc), whereas at each corner of the spectrum, frequency is the highest. As low frequency components are sufficient for image reconstruction, we have set low cutoff frequency value. Values of cutoff frequency have been varied to obtain better result.

For our research work, we have used satellite image which is captured by LISS III 23 m sensor [8, 9]. In red, green, short wave infrared and near-infrared spectral bands, LISS-III which is a well-known sensor, provides good output. In all above-mentioned bands, using only 23 m resolution LISS-III can capture an area of 141 km range.

In this paper, we have used the mathematical formula of Butterworth low-pass filter. Three different cutoff frequencies have been selected for our research work, the output images of each frequency are shown and PSNR value of the output images is measured. Though various techniques were used earlier, the visual quality of their selected satellite images was not good. Whereas in our method, we have used a good quality satellite image captured by LISS-III sensor. The methodology has been discussed in Sect. 2 followed by results in Sect. 3, and conclusion has been drawn in Sect. 4.

2 Methodology

Here, we have used frequency domain compression technique of satellite images. Let us assume that $f(x, y)$ is the original image which is used for compression. For a satellite image of size $M \times N$, the two-dimensional DFT is given by:

$$F(u, v) = \sum_{x=0}^{M-1} \sum_{y=0}^{N-1} f(x, y) \mathrm{e}^{-i2\pi\left(\frac{ux}{M} + \frac{vy}{N}\right)} \tag{1}$$

where $F(u, v)$ is the Fourier transform of the image.

The transfer function of the low-pass filter is

$$\begin{aligned} \mathrm{H}(u, v) &= 1, \quad \text{for } \; D(u, v) \leq D_0 \\ &= 0, \quad \text{for } \; (u, v) > D_0 \end{aligned} \tag{2}$$

Here, cutoff frequency is represented by D_0 and $D(u, v)$ represents the gap of point (u, v) from center point in the frequency band.

Here,

$$D(u, v) = \left[\left(u - \frac{M}{2} \right)^2 + \left(v - \frac{N}{2} \right)^2 \right]^{\frac{1}{2}} \tag{3}$$

and

$$H(u, v) = \frac{1}{\left[1 + \frac{D(u,v)}{D_0} \right]^{2n}} \tag{4}$$

where n is the order of spatial domain frequency.

PSNR value is calculated by

$$\mathrm{MSE} = \frac{1}{mn} \sum_{y=1}^{m} \sum_{x=1}^{n} [f(x, y) - g(x, y)]^2 \tag{5}$$

where MSE = Mean Square Error; $g(x, y)$ = Extracted image; m, n denote dimension of images

$$\mathrm{PSNR} = 20 \log_{10} \frac{255}{\sqrt{\mathrm{MSE}}}$$

3 Result

The proposed technique is based on captured satellite image by LISS III 23 m sensor. The image $f(x, y)$ is shown in Fig. 1a and spectrum of the image is shown in Fig. 1b.

In frequency domain filtering, we have used Butterworth low-pass filter with cutoff frequencies as 150, 200 and 250. We find that below 150, quality of the filtered images is not satisfactory. The output of the filtered images using these above-mentioned cutoff frequencies is shown in Fig. 2a–c. Dimension (no of row and column) of the compressed output images is same with original image (Tables 1 and 2).

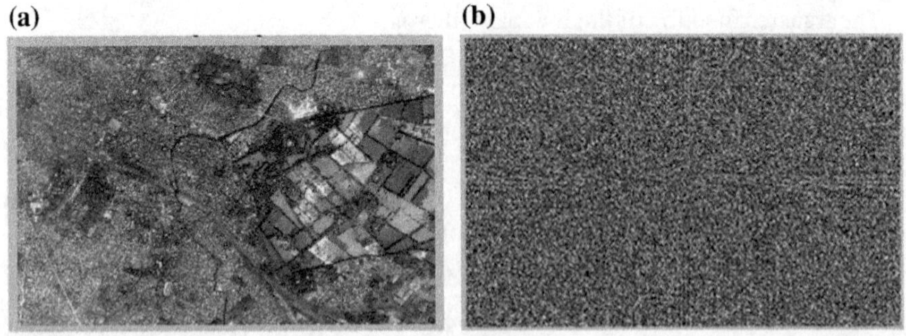

Fig. 1 **a** Image $f(x, y)$, **b** spectrum of the image. Picture Courtesy—ISRO-Bhuvan Web site [The Satellite Image used in this paper is collected from ISRO-Bhuvan Web site. Dr. D. Chakraborty (Sr. Scientist of RRSC–ISRO and co-author of this paper provided this image)]

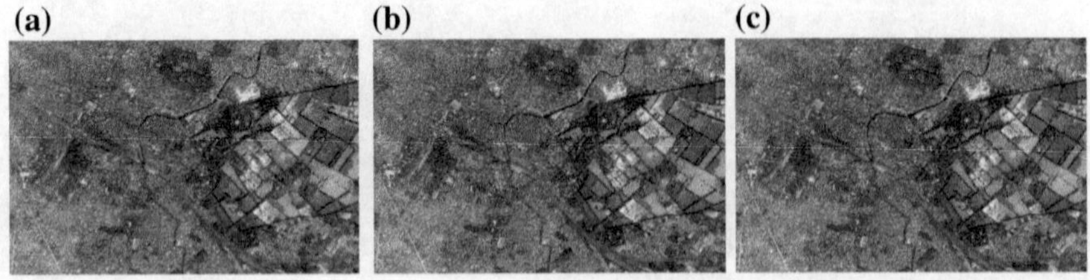

Fig. 2 Image obtained after applying frequency domain filtering with cutoff frequencies **a** 150, **b** 200 and **c** 250

Table 1 PSNR calculation

Image	Size (kB)	PSNR
Original image	166	
Cutoff frequency is 150	103	30.6
Cutoff frequency is 200	121	31.2
Cutoff frequency is 250	136	31.8

Table 2 Energy transmission

Image	Size (kB)	Compression ratio (compressed image/original image)	Energy transmitted (%)
Cutoff value is 150	103	0.62	65
Cutoff value is 200	121	0.73	77
Cutoff value is 250	136	0.82	84

4 Conclusion

A simple technique of satellite image compression in frequency domain using Butterworth low-pass filter is proposed here. During satellite image compression, too much information loss cannot be allowed as a minute part of the image represents a particular geographical area. That is why in our proposed technique, we have not got a very high compression ratio. Compared with earlier methods, visual quality of the selected satellite images is very good as it is captured by optical LISS-III sensor. Moreover, in this method, no post processing of the image is required as the quality of the output images is reasonably good. PSNR value is calculated in each case where it shows that quality of the output images is satisfactory. Percentage of energy transmission from original image has been discussed for comparison. This method can be equally effectively used for color satellite image by applying Butterworth low-pass filter in three different planes of the image which will be discussed in the future communication.

Acknowledgements The Satellite Image used in this paper is collected from ISRO-Bhuvan Web site. Dr. D. Chakraborty (Sr. Scientist of RRSC–ISRO and co-author of this paper provided this image).

References

1. Al Mamun, Md., Ali Hossain, Md.: Satellite image compression using integer wavelet regression. In: International Conference on Electrical, Computer and Communication Engineering (ECCE) (2017). https://doi.org/10.1109/ecace.2017.7912944
2. Sahnoun, K., Benabadji, N.: Satellite image compression technique using noise bit removal and discrete wavelet transform. Int. J. Imaging Robot. **15**(3) (2015)
3. Sahnoun, K., Benabadji, N.: Satellite image compression technique based on the evidence theory. Adv. Comput. Int. J. (ACIJ) **5**(1) (2014)
4. Sahnoun, K., Benabadji, N.: Satellite image compression algorithm based on the FFT. Int. J. Multimedia Appl. (IJMA) **6**(1) (2014)
5. Memane, T., Ruikar, S.D.: Selection of wavelet for satellite image compression using picture quality measures. In: International Conference on Communication and Signal Processing (2014)
6. Susilo, R.M., Bretschneider, T.R.: On the real time satellite image compression of X-Sat (2003). https://doi.org/10.1109/ICICS.2003.1292497
7. Hacihalilogluand, I., Kartal, M.: DCT and wavelet based image compression in satellite images. In: Recent Advances in Space Technologies (RAST '03) (2003)
8. Naidu, Ch.R., Giridhar, M.V.S.S.: Un-supervised classification of rice crop using IRS LISS III satellite images for Wazirabad command area. IJEDR **4**(2) (2016). ISSN: 2321-9939
9. Madhubala, M., Mohan Rao, S.K., Ravindra Babu, G.: Classification of IRS LISS-III images by using artificial neural networks. IJCA Special Issue on "Recent Trends in Image Processing and Pattern Recognition" RTIPPR (2010)

Advanced Wavelet Transform for Image Processing—A Survey

Manas Saha, Mrinal Kanti Naskar and B. N. Chatterji

Abstract Over the last few years, the wavelet transform has played a significant role in the field of multiresolution image analysis. The shortcomings of the wavelet transform laid the foundation of many advanced wavelets. This review paper brings together ten advanced wavelets on a common platform to discuss their importance, concept, architecture, merits and demerits in various fields of image processing. The relationships among the different advanced wavelets are also illustrated here. A comparison table serves as a catalog to know the recent trends and applications of the advanced wavelets.

Keywords Anisotropy · Feature extraction · Multiresolution · Sampling

1 Introduction

The efficient feature extraction and representation have always been a great challenge to the researchers. There has been a long way of evolution from the Fourier Transform, Discrete Fourier Transform, Fast Fourier Transform, Short-Time Fourier Transform to the Wavelet Transform. Some of the applications of the wavelet transform in image processing are coding, fusion, enhancement, compression, denoising, segmentation, content-based image retrieval and so on. Though the wavelet transform is a very popular and promising tool for sparse representation of objects with point singularities, its efficiency is greatly challenged by the objects with line singularities due to the lack of directionality and anisotropy as suggested by Do and

M. Saha (✉)
Siliguri Institute of Technology, Siliguri, WB 734009, India
e-mail: manassaha77@yahoo.com

M. K. Naskar
Jadavpur University, Jadavpur, Kolkata, WB 700032, India
e-mail: mrinaletce@gmail.com

B. N. Chatterji
B. P. Poddar Institute of Management and Technology, Kolkata, WB 700052, India
e-mail: bnchatterji@gmail.com

J. K. Mandal et al. (eds.), *Information, Photonics and Communication*,
Lecture Notes in Networks and Systems 79,
https://doi.org/10.1007/978-981-32-9453-0_19

Vetterli [1]. Therefore, the urgent need of higher-dimensional sparse representation of images motivates the researchers to innovate the mathematical transforms called the advanced wavelets. Here, advanced wavelet is referred as X-let.

The organization of the rest of the paper is given below. Having addressed the significance of X-let in the introductory section, ten different X-lets are presented in Sect. 2. Section 3 exhibits a compact chart for the overall comparison of the X-lets. A brief study on X-let using numerical data obtained from experimental results is mentioned in Sect. 4. The conclusion with the direction of future work is drawn in Sect. 5.

2 Ten Different X-lets

The ten different X-lets namely ridgelet, curvelet, contourlet, wedgelet, bandlet, grouplet, surfacelet, shearlet, Gabor wavelet and brushlet are discussed below.

2.1 Ridgelet Transform

The ridgelet transform was developed by Candes and Donoho [2] in 1999 to break the limitation of the wavelet transform in the higher dimension. The ridgelet analysis is related to the theory of approximation-based superposition of the ridge functions [2]. Reviewing the fundamentals and flowgraph of the ridgelet transform presented by Starck et al. [3], it can be concluded that the ridgelet transform can resolve the problem of sparse approximation of smooth objects with straight edges. Unfortunately, in practice, the edges of the images are typically curved instead of being straight. Therefore, to study the curve singularities, one probable solution is to segregate the image into a number of sub-images and then apply the ridgelet transform individually to the sub-images as discussed by Donoho and Duncan [4, Fig. 2].

2.2 Curvelet Transform

The curvelet transform was pioneered by Candes and Donoho [5] as a new multiscale directional transform. This transform in two dimensions (2D) provides nearly optimal sparse representation of objects having singularities along smooth curves. The first generation of the curvelet was developed in the continuous domain by multiscale filtering and "embedded" ridgelet transform. Unfortunately, the redundancy and the slow nature of the first generation curvelet motivated researchers to develop the second generation curvelet transform [5, 6] which is determined by the frequency partitioning technique without the use of the ridgelet transform, thereby making it a more robust and fast image analysis tool.

2.3 Contourlet Transform

Do and Vetterli [1] reported a double filter bank-based transform called the contourlet transform for obtaining sparse representation of images. This tool provides a variable multiresolution and directional decomposition of images. In the pyramidal filter bank [1, Fig. 7], the first stage is the multiscale decomposition by the Laplacian pyramid (LP) followed by directional decomposition by two-dimensional directional filter bank (2DDFB). The LP captures the point discontinuities and directional filter bank (DFB) links, the point discontinuities into linear structures or contour segments. The LP decomposition generates a bandpass image. This bandpass image so produced is free from "scrambled" frequencies which are associated with wavelet filter.

2.4 Wedgelet, Bandlet and Grouplet Transform

The role and the limitation of the X-lets in resolving the complex image boundaries with typical shapes and conditions have been discussed below with the flowchart given in Fig. 1, where α is the geometrical regularity determining factor. It is known that the geometrically regular functions can be described as a piecewise C^{α}-regular functions (α times continuously differentiable) outside a set of regular edges. But the curvelet transform has the optimal image representation for only $\alpha = 2$. In practice, the images have irregular geometry with either $\alpha < 2$ or $\alpha > 2$ [7]. Therefore, when the question of regularity along the singularities of a surface arises, the failure of the curvelet transform is overcome by a novel approximate scheme called the wedgelet [8]. According to Yang et al. [7], the wedgelet divides the support of the image in dyadic adapted squares. But this approach is suitable for edges without blur and simple geometry images. Therefore, the quest for better X-let continues and one such approximation-based adaptive technique called the bandlet was introduced by Pennec and Mallat [9]. It can suitably capture the geometric regularity along the edges in an image by implementing an adaptive approximation of the image geometry when α is unknown and can fruitfully represent images because each bandlet atom is represented by a geometric flow showing the directions of regular variations of the gray level [7]. The geometry of the bandlet is suitable for analyzing geometrical regular images. But the key drawback of bandlet is that it cannot faithfully represent the complex geometry of textures as that of a wooden fiber. Thus, the researchers came up with another novel approach called the grouplet [10, 7] based on the Gestalt theory [11]. It suggests the recursive use of a set of grouping laws which help to model the edges of images with long range of monotonic turbulent geometry, e.g., wooden texture.

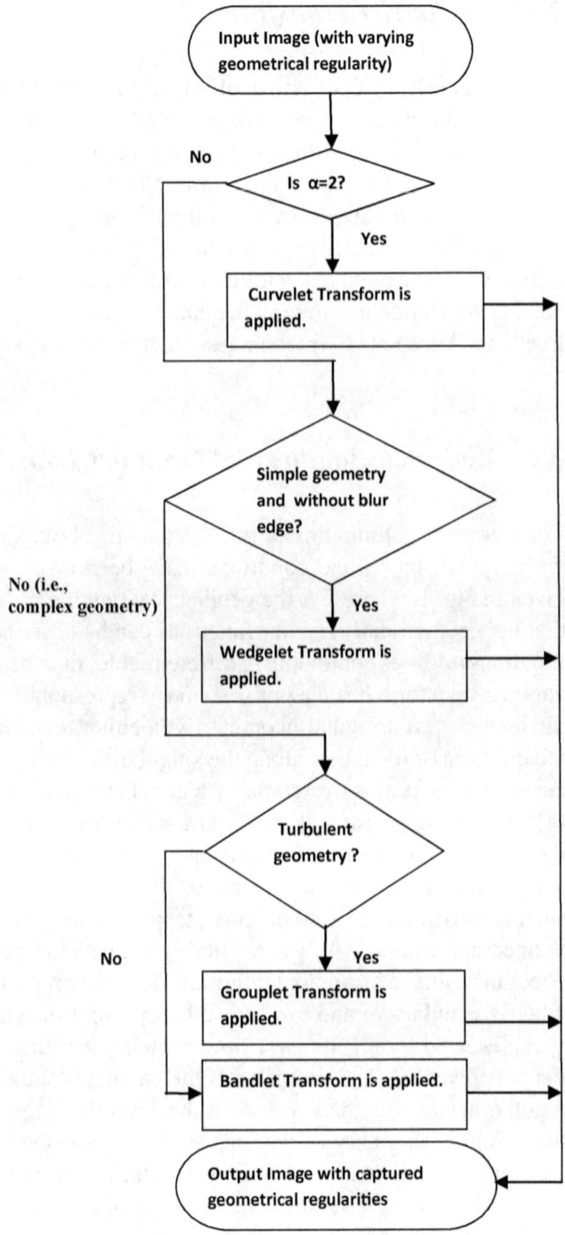

Fig. 1 Flowchart for finding the geometrical regularity of images under different constraints

2.5 *Surfacelet Transform*

The DFB was introduced by Bamberger and Smith in the year 1992 [12]. Lu and Do [13] proposed the extension of 2DDFB to higher dimensions resulting in a new type of filter banks called three-dimensional directional filter bank (3DDFB) which retains the wonderful capacity of directional decomposition of 3D (three dimensional) signals. The 3DDFB, when combined with the LP, gives rise to the 3D surfacelet or surface transform. The block diagram representation of the surfacelet transform is presented by Lu and Do [14]. The surfacelet bears close resemblance with another transform called dual-tree wavelet transform (DTWT). The redundancy ratio in three dimensions is almost the same for both the surfacelet and DTWT. But the advantage of surfacelet over DTWT is that its angular resolution can be refined to a very high level just by raising the number of decomposition levels.

2.6 *Shearlet Transform*

Shearlet introduced by Labate et al. [15] is a new class of multidimensional image representation tool. It is popular due to its ability to represent bivariate functions sparsely. So far various directional transforms like curvelets, contourlets and surfacelets have been addressed for edge representation of images. But none of the X-lets provides a unified treatment of both the continuous and digital setting [16]. This major drawback of these multiresolution methods is overcome by the shearlet due to its uniting capacity of the continuous and digital domain [16]. Another interesting feature of the shearlet is the easy fitting in the framework of affine like systems. Shearlets are basically functions with orientation. The orientation of this function or waveform can be regulated by a parameter called shear parameter. These can be obtained by using dilation, translation and shear transformation of a given function.

2.7 *Gabor Wavelet*

In image processing, the octave-based decomposition of the Fourier plane by the wavelet transform results in a poor angular resolution. But the wavelet packets can decompose the Fourier plane optimally at the cost of the four symmetrical peaks in the frequency plane [17]. Thus, it is difficult to selectively tune and trace a unique frequency. Directionally oriented filter banks, steerable pyramid resolve the random partitioning of the Fourier plane. In this context, the role of Gabor filter needs mentioning. A Gabor filter [18] resembles a wavelet filter bank where an individual filter produces an estimate of the local frequency content. This Gabor filter is a local bandpass filter with joint localization of the spatial and frequency domain. The remarkable feature of Gabor wavelet is that the Gabor basis is not only frequency tunable but also orientation selective. The Gabor wavelets designed in close analogy to visual cortical cells of the mammalian brain helps to decompose a given image into multiple scales and multiple orientations too.

2.8 Brushlet Transform

It is understood from the Gabor filter that the computation load increases abruptly due to the convolution of the original image with so many filters of the bulky Gabor filter bank. Therefore, for better angular resolution, Meyer and Coifman [17] expanded the Fourier plane into windowed Fourier bases to develop a new X-let called brushlet. This is a well-localized complex-valued function bearing a unique peak in frequency domain. Being a complex-valued function, it is associated with a phase giving the knowledge of orientation of the X-let. The size and the location of the brushlets can be adaptively selected in order to obtain the most precise representation of an image in terms of oriented features with all possible directions and low computation load as compared to Gabor filters.

3 Comparison of X-lets

The comparison of the X-lets is provided in Table 1.

4 Short Study on X-lets Using Experimental Results

There are numerous applications which experiment with X-lets. Some of the applications are (i) mammogram denoising using X-let [19, 20]; (ii) wavelet and curvelet based on soft, hard and block thresholding techniques for noise filtration of mammograms [21]. Here, a short study is presented on the numerical data obtained from [21].

The mammograms with Salt and Pepper, Speckle, Poisson, Gaussian noise are denoised by wavelet and curvelet with an intention to exploit the role of underlying soft, hard and block thresholding techniques. The thresholding technique already discussed in [21] refers to either the preservation or elimination of wavelet/curvelet coefficient generated during the implementation of wavelet/curvelet transform to the mammogram image. The average signal to noise ratio (SNR) obtained by implementing soft, hard and block thresholding techniques using wavelet and curvelet transforms to mammograms with different noises are compared in Fig. 2. When the performance of the three thresholding techniques is compared, ignoring the type of transform, block thresholding technique provides a promising result for all noises

Table 1 Comparison of X-lets

X-let	Significant characteristics	Merits	Demerits	Selected applications
Ridgelet	Optimal representation of straight-line singularities	Better than wavelet in attaining low mean square error while representing smooth functions and straight edges	Not suitable for representing images with curves and texture	Space weather monitoring, characterization of nano-structures taken by scanning tunneling microscope
Curvelet	Localized in position, scale and orientation	Better than wavelet and ridgelet in image compression and denoising	Higher computational cost than the wavelet transform especially in higher dimension	Fluid mechanics, Solving of partial differential equations
Contourlet	Direct determination of contourlet transforms on rectangular grid and compactly supported frame	Better than curvelet since immune to quantization noise and noise due to thresholding effect	Implicit oversampling by LP	CBIR, Image representation, denoising and compression
Wedgelet	Sub-band decomposition of the original image followed by the use of the prediction and update operator	Better than wavelet in representing the linear features of an image	Application limited to the horizon class image, i.e., binary image with single edge	Image enhancement and denoising
Bandlet	Specialized in capturing geometric regularity along the edges of an image	Better than curvelet and wedgelet in representing the edges of images having complex geometry like boiling water image	Unfaithful representation of the complex geometry of textures	Image compression, restoration and denoising

(continued)

Table 1 (continued)

X-let	Significant characteristics	Merits	Demerits	Selected applications
Grouplet	The length and width of the individual atom can be modulated to match the different types of geometrical structures of natural images	Better than curvelet, bandlet and wedgelet in representing the long range regularity of fine elongated structures like hair texture	Difficult to adjust the complexity of the association fields for coding it with less number of bits for getting satisfactory compression result	Image in painting, texture analysis
Surfacelet	Capture and represent the signal singularities lying on smooth surfaces noticed in 3D medical and video images	Better than dual tree wavelet transform due to high degree of angular resolution caused by increasing number of decomposition levels	High redundancy of surfacelet is reduced in the higher dimensions	Processing of multidimensional volumetric data like seismic imaging and video clips, denoising of video signals
Shearlet	Sparse representation of bivariate functions	Better than classical wavelet because it gives information regarding the directionality within the image	Being associated with scaling and translation parameter it cannot detect directionality like conventional wavelet transform	Representation of signal at higher dimension
Gabor Wavelet	Multiple scale and orientation decomposition of an image	Gabor basis is frequency tunable and orientation selective	Difficult to reconstruct the Gabor wavelets by linear superposition due to the unavailability of orthonormal bases	Bio-metric application like iris identification, face recognition
Brushlet	Localized complex function with one peak in the frequency domain	Feature retrieval rate better than Gabor wavelets due to less computation	Though this X-let is based on adaptive representation but its construction is hard to implement	Texture segmentation, retrieval and classification

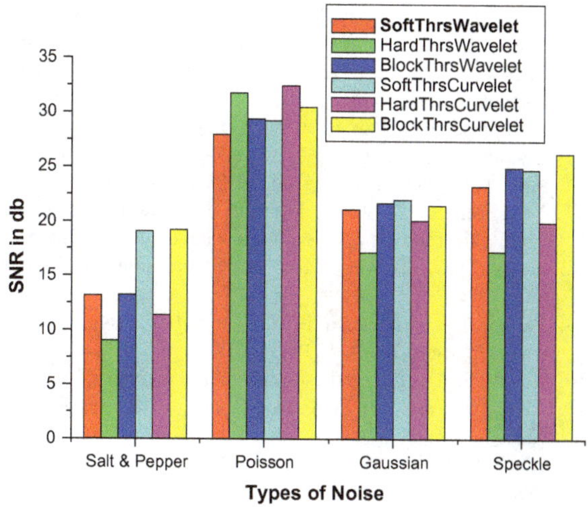

Fig. 2 Comparison of mean SNR found using soft, hard and block thresholding techniques based on wavelet and curvelet transforms [21]

except Poisson noise. Hard thresholding gives satisfactory result for Poisson noise. When the overall denoising performance is the primary issue, curvelet outperforms wavelet.

5 Conclusion

The primary goal of understanding the significance and classification of X-let has been consistently presented in this paper. This paper describes the shortcomings of an X-let and its gradual development. One observation about the X-lets is that despite sharing a common structure or a similar function, they are very much application specific. And this paper gives us an idea of the applicability of the different members of the X-let family for various image processing purposes. The several less explored areas like the implementation of curvelet transform in seismic imaging or the upconversion of ordinary television image into high definition image using bandlet can be the directions of the future work.

References

1. Do, M.N., Vetterli, M.: The contourlet transform: an efficient directional multiresolution image representation. IEEE Trans. Image Process. **14**(12), 2091–2106 (2005)
2. Candès, E.J., Donoho, D.L.: Ridgelets: a key to higher-dimensional Intermittency? Philos. Trans. A Roy. Soc. Lond. **357**(1760), 2495–2509 (1999)
3. Starck, J.L., Candes, E.J., Donoho, D.L.: The curvelet transform for image denoising. IEEE Trans. Image Process. **11**(6), 670–684 (2002)

4. Donoho, D.L., Duncan, M.R.: Digital curvelet transform: strategy, implementation and experiments. In: Proceedings of Aerosense 2000, Wavelets Applications VII, vol. 4056, pp. 12–29. SPIE. Available http://www.curvelet.org/papers/DCvT99.pdf (2000)
5. Candès, E.J., Donoho, D.L.: New tight frames of curvelets and optimal representation of objects with piecewise C^2 singularities. Commun. Pure Appl. Math. **57**(2), 219–266 (2004)
6. Candes, E., Donoho, D.: Continuous curvelet transform. I. Resolution of the wavefront set. Appl. Comput. Harmon. Anal. **19**(2), 162–197 (2005)
7. Yang, S., Lu, Y., Wang, M., Jiao, L.: Low bit rate SAR image coding based on adaptive multiscale bandlets and cooperative decision. Sig. Process. **89**(10), 1910–1920 (2009)
8. Donoho, D.L.: Wedgelets: nearly minimax estimation of edges. Ann. Stat. **27**(3), 859–897(1997)
9. Pennec, E.L., Mallat, S.: Sparse geometric image representations with bandlets. IEEE Trans. Image Process. **14**(4), 423–438 (2005)
10. Mallat, S.: Geometrical grouplets. Appl. Comput. Harmon. Anal. **26**(2), 161–180 (2009)
11. Ellis, W.D. (ed.): A Source Book of Gestalt Psycology. The Gestalt Journal Press Inc., US (2007)
12. Bamberger, R.H., Smith, M.J.T.: A filter bank for the directional decomposition of images: theory and design. IEEE Trans. Sig. Process. **40**(4), 882–893 (1992)
13. Lu, Y.M., Do, M.N.: Multidimensional directional filter banks and surfacelets. IEEE Trans. Image Process. **16**, 918–931 (2007)
14. Lu, Y., Do, M.N.: 3-D directional filter banks and surfacelets. In: Proceedings of SPIE Conference on Wavelet Applications in Signal and Image Processing XI, San Diego, USA (2005)
15. Labate, D., Kutyniok, G., Lim, W.Q., Weiss, G.: Sparse multidimensional representation using shearlets. In: Proceedings SPIE, Wavelets XI, 59140U (2005). https://doi.org/10.1117/12.613494
16. Han, B., Kutyniok, G., Shen, Z.: Adaptive multiresolution analysis structures and shearlet systems. SIAM J. Numer. Anal. **49**, 1921–1946 (2011)
17. Meyer, F.G., Coifman, R.R.: Brushlet: a tool for directional image analysis and image compression. Appl. Comput. Harmon. Anal. **4**, 147–187 (1997)
18. Jain, A.K., Farrokhnia, F.: Unsupervised texture segmentation using Gabor filters. Pattern Recognit. **24**(12), 1167–1186 (1991)
19. Saha, M., Naskar, M.K., Chatterji, B.N.: Poisson noise removal from mammogram using Poisson unbiased risk estimation technique. In: Proceedings of Second International Conference on Information Systems Design and Intelligent Applications—2015 (INDIA 2015), Kalyani, West Bengal, 8–9 Jan 2015, vol. 2, pp. 327–335. https://doi.org/10.1007/978-81-322-2247-7_3
20. Saha, M., Naskar, M.K., Chatterji, B.N.: Mammogram denoising by curvelet transform based on the information of neighbouring coefficient. In: Proceedings of 2015 Third International Conference on Computer, Communication, Control and Information Technology (C3IT), Hoogly, West Bengal, 7–8 Feb 2015, pp. 1–6. Available at IEEE Xplore Digital Library. https://doi.org/10.1109/c3it.2015.7060180
21. Saha, M., Naskar, M.K., Chatterji, B.N.: Soft, hard and block thresholding techniques for denoising of mammogram images. IETE J. Res. **61**(2), 186–191 (2015)

Classification of RBC and WBC in Noisy Microscopic Images of Blood Smear

Sayantari Ghosh and Saumik Bhattacharya

Abstract Blood cell counts are one of the vital signs that play a major role in medical diagnostics. The process of estimating complete blood count (CBC) is either expensive or time taking depending on the technique that is used for counting. A manual differential technique, that is often used to get authentic count and abnormalities present in a stained blood smear, requires a lot of human expertise and time. Dedicated efforts are given to make the classification and counting process of blood profiles automatic and robust. In this paper, first, we build a neural network-based multiclass classifier that can classify between RBC, WBC, and platelets. However, we demonstrate that the presence of a small amount of random noise can drastically decrease the performance of the designed classifier. Thus, we design a preprocessing network to enhance the quality of the input image so that the classifier can faithfully classify the blood cells even in adverse imaging conditions.

Keywords RBC · WBC · Enhancement · Denoising

1 Introduction

Blood cell counting is one of the essential tests that are often carried out to detect infection, allergies, disorders, anemia, and many other diseases. Complete blood count (CBC) contains a series of tests that majorly include WBC and RBC counts. Further diagnosis of the patient also requires a thorough analysis of RBC, WBC, and platelet cells. Though, there is some equipment available that performs counting and the analyzes automatically, most of the authentic analyzes are done manually. Manual processing of stained blood smear is time taking and difficult. Thus, extensive researches are going on to make the entire process automatic and fast. In [1], the authors used background suppression followed by color features extraction to classify

---

S. Ghosh
National Institute of Technology Durgapur, Durgapur, India

S. Bhattacharya (✉)
Indian Institute of Technology Roorkee, Roorkee, India
e-mail: saumikfec@iitr.ac.in

J. K. Mandal et al. (eds.), *Information, Photonics and Communication*,
Lecture Notes in Networks and Systems 79,
https://doi.org/10.1007/978-981-32-9453-0_20

WBC using SVM. A non-parametric mean shift algorithm is proposed by Comaniciu and Meer [2] to automatically segment WBC in a blood smear. In [3], scale-space filtering and watershed clustering are used to segment WBC. Liao and Deng [4] used regional information of binary images to detect and count cells in a blood smear. A model-based combinatorial algorithm is used in [5] to segment WBC. Ongun et al. [6] proposed an automated differential blood counting system using active contour models and different classifiers. In [7], the authors used k-mean clustering followed by EM algorithm to segment and count WBC. In [8], the authors enhanced the green channel and contour-based detection to segment erythrocytes and leukocytes in a blood smear. Poostchi et al. [9] combined texture feature with RGB feature to design LoG-based automatic RBC counter. In [10], the author exhibits that direct least-squares fitting is an efficient way to analyze RBC in a blood smear. Dvanesh et al. [11] combined morphological operations and circular Hough transform to count the number of blood cells.

In this paper, we first propose a convolutional neural network (CNN) that classifies RBC, WBC, and platelet. However, we demonstrate that the performance of the classifier degrades as the quality of the input image becomes worse. It is important to have a robust classifier for this particular task as the design of an automated blood cell counting system may have different lighting conditions and imaging devices [12]. Thus, we design another fully convolutional network (FCN) that restores the blood cell images and helps the multiclass classifier to correctly recognize the blood cells.

2 Proposed Method

In this paper, we propose two different convolutional neural network models that enhance and classify input images of blood cells. To appreciate the necessity of the enhancement network, we first discuss the multiclass classification network and exhibit its limitations. We will further discuss the enhancement network and will demonstrate the advantages of the preprocessing unit.

2.1 *Multiclass Blood Cell Classifier*

For a given color image of a blood cell, we resize the image to 64×64 dimensional color image and normalize it accordingly. The classifier contains three convolutional layers with 3×3 filters, stride 1. The first two convolutional layers contain 32 filters, whereas the last convolutional layer contains 64 filters. We use *ReLU* activation for each convolutional layer. We introduce max-pooling of size 2 after each convolutional layer to reduce data dimension and to impose the learning of invariance. We convert the output of the third convolutional layer to a column vector and connect that to a *ReLU* activated dense layer of size 64. We finally add another dense layer of size

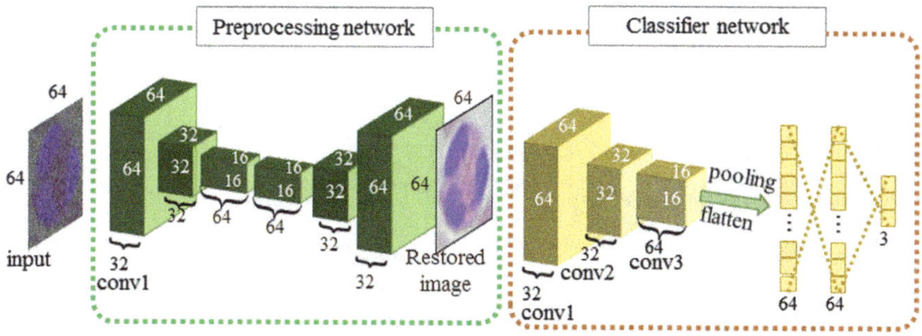

Fig. 1 Block diagram of the proposed preprocessing network (left block) and the classifier network (right block). The preprocessing network takes the noisy input image and gives the restored image as output. The restored image acts as input to the classifier

3 which has *SoftMax* activation as our final output layer. We use 50% dropout in between the dense layers as regularizer to avoid overfitting. We train the network by minimizing the categorical cross-entropy loss using ADAM optimizer. The classifier architecture is shown in Fig. 1. We train the classifier with 6400 blood cell images and validate on 3200 images. The dataset is publicly available in the link.[1] The proposed classifier achieves 98.4% validation accuracy for noise-free input images.

Though we achieve a very high validation accuracy using the proposed network, we further investigate the robustness of the architecture. As dedicated efforts are going on to make the automated blood cell counting systems user-friendly and cheap, it is important that the architecture must be invariant to different image degradations that may appear due to lack of expertise, poor imaging condition, or cheap imaging devices [13]. We apply Gaussian blur kernel with $\sigma \in [0, 35]$ to understand the effect of blurring on the network. We observe that Gaussian blurring does not have any significant effect on the classification accuracy. The change of accuracy in the presence of Gaussian blur is shown in Fig. 2a.

However, experiments show that the contrast and the presence of noise drastically change the validation accuracy. As shown in Fig. 2b, the presence of noise, a small variation in global illumination significantly degrades the overall performance of the system. The intuitive reason behind this degradation could be the dataset that has been used to train the network. As the dataset contains only good quality images of blood cells, it is difficult to have a robust representation of the input in the presence of noise under low illumination. One way to solve the problem is to train the classifier with some degraded input data along with the good quality cell images. Though this solution may provide better robustness than the previously trained model, we focus to develop a preprocessing network that enhances an input image by correcting the contrast and removing the noise. The main motivation behind the preprocessing step is to generate intermediate noise-free images that not only improves the classification accuracy in the presence of degradation, but also helps in manual analysis for any disorder or in cross-checking. We discuss the proposed preprocessing network below.

[1] https://www.kaggle.com/paultimothymooney/blood-cells.

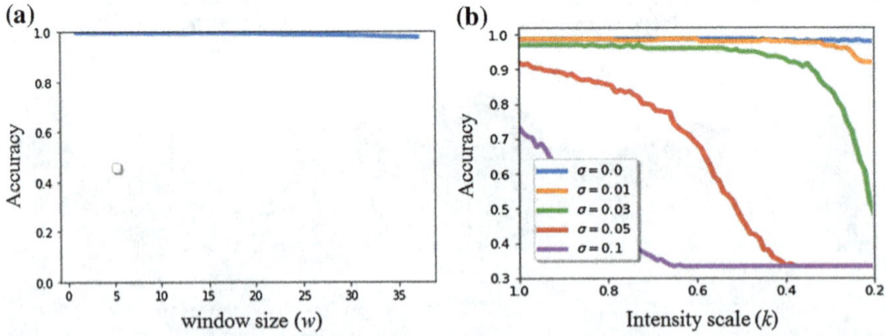

Fig. 2 Effect of image degradations on classifier network when the classifier is trained with noise-free images and tested: **a** in the presence of Gaussian blur with window size w; **b** in the presence of global intensity scaling and AWGN noise with standard deviation σ

2.2 *Preprocessing Network*

To enhance the quality of a blood cell image, we design a fully convolutional network (FCN) that takes noisy image as input and generates denoised contrast-enhanced image as output. The reason behind selecting FCN architecture is that it is not only faster to train, but also has lesser number of parameters compared to any dense layer-based architecture that produces reconstructed image with the same dimension. The proposed network has a convolutional input layer with 32 filters of dimension 3×3. Next, we have two *ReLU* activated convolutional layers with 32 and 64 filters of dimension 3×3, respectively, with stride 2. These two layers produce a compact representation of the input image in higher-dimensional space. We add another convolutional layer with 32 filters of dimension 3×3 and stride 1. The expanding part of the network contains two different convolutional blocks which individually contains upsampling by 2 and *ReLU* activated convolutional layer with 32 filters of dimension 3×3 and stride 1. The output of the network is taken from a convolutional layer with 3 filters of dimension 3×3 and stride 1. The network is trained using ADAM optimizer to minimize mean absolute error of the reconstructed denoised image and the actual ground-truth images.

3 Results

The entire classification architecture with the preprocessing unit is shown in Fig. 1. Instead of an end-to-end training, both the units are trained separately to ensure that the intermediate restored image is the denoised version of the noisy input image for further manual analysis. The preprocessing network has total 66,467 numbers of parameters, and the classification network has total 176,355 numbers of parameters.

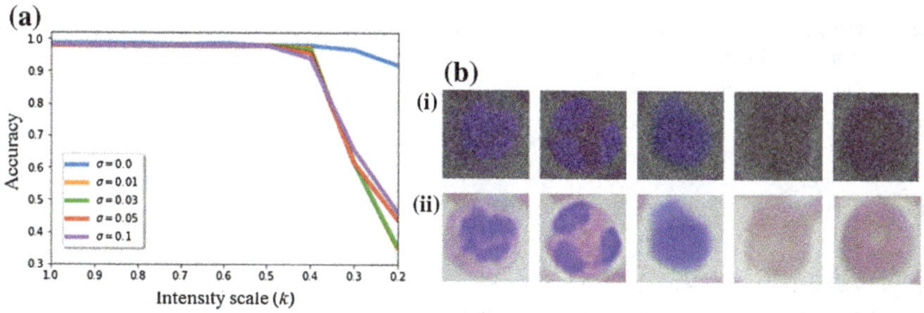

Fig. 3 Performance of the classifier in the presence of the preprocessing unit: **a** change in accuracy when the global contrast is scaled and AWGN is present; **b** noisy input images of the blood cells (upper row), and respective restored color images (bottom row) using preprocessing network

As shown in Fig. 2b, the classifier performs satisfactorily for noise-free images. However, when the global illumination of the input images is scaled with a scaling parameter k, and an additive white Gaussian noise (AWGN) with standard deviation σ is applied, the performance of the classifier degrades significantly. We select the global illumination scaling as it represents the loss of contrast that often happens in poor lighting condition, and the noise, that is produced by any cheap image sensor, is approximated by AWGN in this work. In Fig. 3a, it can be observed that the classifier becomes much robust when the preprocessing network is coupled as shown in Fig. 1. To interpret the reason behind this robustness, we check the intermediate denoised image, and as shown in Fig. 3b, we observe that the preprocessing network denoises the noisy color image and rectifies the contrast information preserving the structural information of the input. When we degrade the 3200 images that we use to validate our networks with a scaling factor k randomly chosen from a uniform distribution from (0, 1], and with AWGN of 0 mean and σ randomly chosen from a uniform distribution between (0, 0.1), we achieve average structural similarity measure (ASSIM) of 0.98.

4 Conclusion

Classification of noisy data is known to be a difficult, yet important problem as it inspires robust solutions that can be applied in real-world scenarios. In this paper, we address the classification problem of noisy blood cell data with a motivation to make the visual automatic blood cell counting process more robust. We show that a classifier trained with noise-free well-contrast images perform poorly under minor image degradations. The proposed preprocessing network not only helps to denoise any input blood cell image, but also provides robustness to the classifier to achieve satisfactory classification accuracy under prominent image degradation. In the future, it is necessary to design preprocessing modules that can provide robustness to the blood cell localization task under different image conditions.

Acknowledgements Authors would like to thank Dr. Shahana Gupta, Calcutta Medical College for data and valuable comments and discussions.

References

1. Ramoser, H., et al.: Leukocyte segmentation and classification in blood-smear images. In: 27th Annual International Conference of the Engineering in Medicine and Biology Society, 2005 (IEEE-EMBS 2005). IEEE (2006)
2. Comaniciu, D., Meer, P.: Cell image segmentation for diagnostic pathology. In: Advanced Algorithmic Approaches to Medical Image Segmentation, pp. 541–558. Springer, London (2002)
3. Jiang, K., Liao, Q.-M., Dai, S.-Y.: A novel white blood cell segmentation scheme using scale-space filtering and watershed clustering. In: 2003 International Conference on Machine Learning and Cybernetics, vol. 5. IEEE (2003)
4. Liao, Q., Deng, Y.: An accurate segmentation method for white blood cell images. In: Proceedings. 2002 IEEE International Symposium on Biomedical Imaging. IEEE (2002)
5. Nilsson, B., Heyden, A.: Model-based segmentation of leukocytes clusters. In: Proceedings. 16th International Conference on Pattern Recognition, vol. 1. IEEE (2002)
6. Ongun, G., et al.: An automated differential blood count system. In: Engineering in Medicine and Biology Society, 2001. Proceedings of the 23rd Annual International Conference of the IEEE, vol. 3. IEEE (2001)
7. Sinha, N., Ramakrishnan, A.G.: Automation of differential blood count. In: TENCON 2003. Conference on Convergent Technologies for the Asia-Pacific Region, vol. 2. IEEE (2003)
8. Ismail, B., Moetesum, M.: Automated detection and quantification of erythrocytes and leukocytes from Giemsa stains of blood smear. In: 2018 14th International Conference on Emerging Technologies (ICET). IEEE (2018)
9. Poostchi, M., et al.: Malaria parasite detection and cell counting for human and mouse using thin blood smear microscopy. J. Med. Imaging **5**(4), 044506 (2018)
10. Abu-Qasmieh, I.: Novel and efficient approach for automated separation, segmentation, and detection of overlapped elliptical red blood cells. Pattern Recognit. Image Anal. **28**(4), 792–804 (2018)
11. Dvanesh, V.D., et al.: Blood cell count using digital image processing. In: 2018 International Conference on Current Trends towards Converging Technologies (ICCTCT). IEEE (2018)
12. Habibzadeh, M., et al.: Counting of RBCs and WBCs in noisy normal blood smear microscopic images. In: Medical Imaging 2011: Computer-Aided Diagnosis, vol. 7963. International Society for Optics and Photonics (2011)
13. Roy, P., et al.: Effects of Degradations on Deep Neural Network Architectures. arXiv preprint arXiv:1807.10108 (2018)

Impact of Curvature on Intensity-Based Non-rigid Medical Image Registration

Prasenjit Kumar Mudi

Abstract The non-rigid registration problem has gained much prominence in the area of medical image analysis. Graph cut-based solution to such a non-rigid registration problem is shown to improve the accuracy of the registration results over other existing approaches. Incorporation of curvature information in graph cut-based non-rigid bio-medical image registration has not been explored yet. In this work, a novel approach based on curvature is proposed to improve the performance of the Graph-Cut solution. In this method, first of its kind, the impact of curvature in an intensity-based non-rigid registration is well demonstrated. At first, the extraction of mean and Gaussian curvatures from the brain MRI are outlined and shown appropriate results. Based on the mismatch pattern in curvature distribution, a local displacement map (LDM) is generated for every such unmatched pixel in source image. The map essentially indicates the local displacement such as source pixel should undergo within a local neighborhood to achieve best curvature-wise matching with the corresponding target pixel. Experimental results clearly corroborate to the fact that intensity-based registration along with curvature can yield better registration accuracy.

Keywords Mean curvature · Gaussian curvature · Non-rigid registration · Surface class · Local displacement map

1 Introduction

The problem of image registration arises when images of an object are captured, for example, at different times, orientations, and/or different imaging devices. The primary motive is to combine information and to integrate useful data from different images. However due to spatial distortion (e.g., introduction of motion), same object

P. K. Mudi (✉)
Department of Electronics and Communication Engineering, BP Poddar Institute of Management and Technology, Kolkata 700052, India
e-mail: pmudi.ju26@gmail.com

J. K. Mandal et al. (eds.), *Information, Photonics and Communication*,
Lecture Notes in Networks and Systems 79,
https://doi.org/10.1007/978-981-32-9453-0_21

therefore cannot directly be comparable. The image registration does geometrically transform images to minimize such distortions.

In medical domain, for example, to detect brain tumors, doctors often get misdirected after observing scanned brain slice images obtained from different imaging devices [1]. Such erroneous results can lead to wrong treatments for the patients. Detection of such brain tumors demands very accurate registration which can be possible if curvature information is introduced in existing intensity-based greater level accurate Graph-Cut-based non-rigid image registration technique [2]. Another example with diabetic retinopathy, an eye disease, which occurs due to diabetes, can eventually lead to blindness. Detection of such disease by analyzing retinal fundus images also requires high accurate image registration [3].

1.1 Concept of Curvature

From geometry, the second-order derivative of a two-dimensional curvilinear function $y = f(x)$ i.e. $\frac{d^2y}{dx^2}$ represents the curvature of a point on the surface. In Fig. 1, at each point on the curve, the direction of tangent bends at a certain rate in different directions. A measure of this rate of bending is called curvature at that point and thus defined as [4]:

$$\kappa = \left| \frac{f''(x)}{\left[1 + (f'(x))^2\right]^{3/2}} \right| \tag{1}$$

Now, the curvature information of any pixel in a digital image can be mathematically represented by two principal curvatures, namely the mean curvature and Gaussian curvature as will be discussed in later section. For image registration with non-rigid geometry, two corresponding pixels in source and target images will more accurately be registered if their curvature properties are also considered along with their intensity maps.

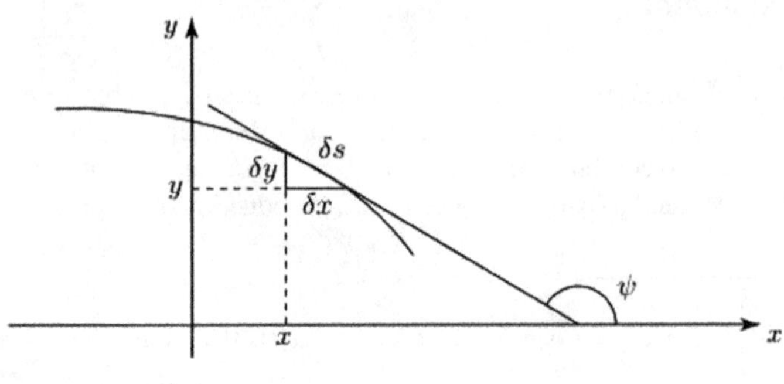

Fig. 1 Definition of curvature [4]

1.2 Non-rigid Registration

Non-rigid image registration has a wide acceptance in biomedical image analysis. The process of image registration is to align different images in a single coordinate system to perform the comparison of features between corresponding pixels of the images. The images may be captured at different instants or using different imaging modalities.

Generally, image registration is of two types: rigid registration and non-rigid registration. In rigid registration, the images are subjected to global transformation such as translation, rotation, or both with respect to one image to the other to achieve correspondence. In sharp contrast, non-rigid image registration, some localized transformation (that may be confined to local neighborhood of a pixel, e.g., displacement) is applied to the regions of two images to establish correspondence.

The non-rigid registration has been modeled as an energy minimization problem [5]. This optimization has been done using Graph-Cut algorithms via alpha-expansions [6]. The energy function of this Graph-Cut (GC) method incorporates Sum of Absolute Difference (SAD), Sum of Squared Difference (SSD), or Mutual Information (MI) as similar measures which are all intensity functions [5]. All of these techniques are used to improve the accuracy of registration outputs as per high requirement in patient diagnosis. Naturally, one objective is to further improvise the accuracy of such methods. Other objective is to incorporate higher-order term in the energy function representing surface properties of registered images. Though there can be other motives, we shall focus on the second objective.

In this paper, curvature extraction is depicted in Sect. 3 followed by proposed method in Sect. 4. Finally, some experimental results are shown in Sect. 5.

1.3 Why Curvature in Non-rigid Registration?

Due to an increasing demand for registration accuracy in medical imaging, production of highly accurate results by using graph-cut methods, which are based on intensity functions, always becomes a challenging task. In the existing graph-cut methods, the great accuracy indicates that most of the pixels are correctly registered based on intensity profile. The remaining, though few in numbers, are being compromised. These can be rectified by incorporating higher-order terms or terms contributing to regularization. On the other way, curvature is a local property of a pixel in its neighborhood and is a second-order differential of intensity. So, curvature is a natural choice of selection in the non-rigid registration techniques.

2 Relevant Study

Image registration is a spatial transformation subjected to a set of images in one coordinate system to find the relation between corresponding pixels in the images. One of the scanned images is called as source image or reference image, and the other images are called target images. An image registration algorithm consists of the following modules [5]:

- The measurement metric that represents similarity or dissimilarity of intensity distribution between two images.
- The transformation model representing the input force applied to target image for alignment.
- The optimization to determine the optimum value of the input force required for transformation.

2.1 Intensity-Based Registration

Intensity-based registration utilizes some intensity functions such as SSD and SAD representing the similarity of intensity patterns between images and hence finds the optimal transformation parameter from the energy function based upon the similarity metric. These result in most similar images after correct optimal registration.

2.1.1 Measures of Similarity

Some of the widely used similarity measures are detailed below:

Sum of Squared Difference (SSD)

It is defined as:

$$\mathrm{SSD} = \frac{1}{N}\sum_{X}[T(x) - S(t(x))]^2 \tag{2}$$

It may be sensitive to number of voxels, N, if their intensity differences are large [5].

Correlation Coefficient (CC)

It is defined as:

$$\mathrm{CC} = \frac{\sum_X \left(T(x) - \overline{T}\right).\left(S(t(x)) - \overline{S}\right)}{\sqrt{\sum_X \left(T(x) - \overline{T}\right)^2 . \sum_x \left(S(t(x)) - \overline{S}\right)^2}} \tag{3}$$

SSD and CC are used in images acquired from single modality [5].

Sum of Absolute Difference (SAD)

It is defined as:

$$\mathrm{SAD} = \sum_{x \in X} \|T(x) - S(T(x)\| \tag{4}$$

Advantage of SAD will be discussed in the optimization subSect. 2.1.3.

In all these cases, $T(x)$ is the intensity of a pixel x in target image, T. $S(T(x))$ is the intensity of the corresponding pixel in the transformed source image, $S(T)$.

2.1.2 Transformation Model

The method of image registration spatially transforms a target image, say I_f to align with the source image, say I_r. The transformation T^* can be modeled as [5]:

$$T^* = \arg\min_T C\left(I_r, T\left(I_f\right)\right) + \lambda S(T) \tag{5}$$

If the transformation T is represented by displacement vector, $\boldsymbol{D}$, Eq. (5) can be written as:

$$D^* = \arg\min_D C\left(I_r, D\left(I_f\right)\right) + \lambda S(D) \tag{6}$$

where $D\left(I_f\right)$ denotes the transformed floating image. $S(D)$ is called the smoothness cost, and λ is a constant of penalty to ensure label consistency.

2.1.3 Optimization

As discussed in earlier subSect. 2.1.2 that the transformation as a minimization problem is given by Eq. (6). The first term of the energy function Eq. (6) is data cost, and the second term is called smoothness cost.

Smoothness Term
Data Term

A novel data term has been proposed by Chowdhury et al. [2] as:

$$C\left(I_r, D\left(I_f\right)\right) = \left\lfloor \exp\left(\sum_{x \in X} \left\| I_r - D\left(I_f\right) \right\|\right) \right\rfloor \tag{7}$$

The SAD is directly used in Eq. (7) as a of measure intensity dissimilarity. Firstly, using Eq. (7), the labels can be directly assigned from the corresponding values of C. Secondly, it assigns low labels to the pixels in target image having small intensity differences with respect to the corresponding pixels in source image due to the exponential term. The intensity pattern of images may get shifted by illumination effect. As a result, registration accuracy gets adversely affected. So, Eq. (7) is modified as [2]:

$$C\left(I_r, D\left(I_f\right)\right) = \left\lfloor \exp\left(\sum_{x \in X} \left\| I_r - D\left(I_f\right) \right\| - \partial\left(I_r, D\left(I_f\right)\right)\right) \right\rfloor \tag{8}$$

$$\text{where } \partial\left(I_r, D\left(I_f\right)\right) = \min\left(\sum_{x \in X} \left\| I_r - D\left(I_f\right) \right\|\right) \tag{9}$$

In Eq. (9), X is the local neighborhood of a pixel x in the target image.

Smoothness Term

The smoothness term is proposed [2] to be:

$$S(D) = \sum_{x \in X, y \in N(x)} \| D(x) - D(y) \| \tag{10}$$

where $N(x)$ is the local neighborhood of the pixel x in image X.

The energy function Eq. (6) is analogous to that of cost function of the graph-cut method given by:

$$E_{GC} = \sum_{x \in X} C_x(I_r) + \sum_{x \in X, y \in N(x)} S_{xy}\left(I_x, I_y\right) \tag{11}$$

where C_x is the data term, S_{xy} is the smoothness term and $I_x\left(I_y\right)$ is the intensity of the pixel x (y) [7].

2.2 *Shortcomings of Intensity-Based Registration*

In the preceding section, it is shown how non-rigid registration is being carried out based on intensity pattern. But, a very important shortcoming of this technique is not incorporating curvature information of pixels in an image.

2.2.1 Registration Accuracy

The errors in intensity-based registration can be rectified by using curvature as it will be illustrated in next section the behavior of curvature of every pixel in image. It may happen in some cases that two corresponding pixels are having, for example, flat surface and ridge surface (based on their local surface behavior), but their intensities are different. So, these are the cases of possible improvement of registration accuracy.

2.2.2 Adaptive Window Size

In the earlier work [2], a fixed window size 31×31 has been used to determine how much displacement of pixels take place through this window. But the displacements of pixels are not uniform throughout the images. Therefore, an adaptive window size should be introduced to fix dynamically the window size according to the displacements of pixels in different regions of image.

3 Curvature Extraction

From differential geometries, normal curvature K at some point P on a surface has different values along different directions. The directions along which K have maximum or minimum values are the principal directions at P. The maximum or minimum values of the curvature K are called principal curvatures. These curvature values and the directions at P hence determine the amount of bending of the surface along the principal directions [8].

Earlier, it has been successfully incorporated to detect some surface-related features such as valley surface, flat surface, or ridge surface from images [9]. Such features characterize the maximal curvature at a point on the image surface. Curvature information of any point on a surface can be mathematically modeled with two matrices: first fundamental form matrix and second fundamental form matrix.

3.1 Principle Curvatures

From differential geometries [9], the first fundamental matrix, I_1, is obtained as:

$$I = \begin{pmatrix} E & F \\ F & G \end{pmatrix} \quad \text{so that} |I| = EG - F^2 \tag{12}$$

E, *F*, and *G* are coefficients of matrix *I*.

The second fundamental form matrix is obtained as:

$$II = \begin{pmatrix} L & M \\ M & N \end{pmatrix} \quad \text{so that} |II| = LN - M^2 \tag{13}$$

L, *M*, and *N* are coefficients of matrix *II*.

A quadratic equation is formed as:

$$|I|k_n^2 - (EN + LG - 2FM)k_n + |II| = 0 \,;\, n = 1, 2 \tag{14}$$

where k_1, k_2 are the two principal curvatures of a point on a surface.

Mean curvature of a point is given by:

$$H = (k_1 + k_2)/2 \tag{15}$$

where k_1, k_2 are the roots of Eq. (14).

Gaussian curvature of a point is defined as:

$$K = k_1 k_2 \tag{16}$$

where k_1, k_2 are the roots of Eq. (14).

In this way, for 2D and 3D digital images, every pixel on images can also be characterized by its local principal curvatures: Mean and Gaussian curvatures (*H*, *K*). Here, for non-rigid image registration, (*H*, *K*) curvature extraction is done for brain MRI images few of which are shown in Figs. 4 and 5.

3.2 Surface Primitives

From Sect. 3.1, we have learned to extract the mean and Gaussian curvature of every pixel of an image. These curvature values of every pixel determine how much the local surface of a pixel is bent or curved along with its neighbor pixels. Based on the (*H*, *K*) value being positive or negative or zero, pixel in an image surface can also be characterized by some local surface primitives or labels as given in Table 1 [10].

Table 1 Local surface primitives based on (H, K)

	$H < 0$	$H = 0$	$H > 0$
$K < 0$	Saddle ridge	Minimal surface	Saddle valley
$K = 0$	Ridge surface	Flat surface	Valley surface
$K > 0$	Peak surface	None	Pit surface

From Table 1, one can have necessary information about the local geometry of a pixel. Therefore, in non-rigid image registration, two corresponding pixels between source and target images will be curvature-wise aligned if their local surface primitives are properly matched. It may happen that two corresponding pixels are matched intensity-wise but their local surface primitives are not the same.

4 Proposed Method

As discussed in earlier sections, curvature information should be incorporated along with intensity to achieve better registration accuracy. As curvature is a differential function of intensity, the registration method [2] should yield similar curvature patterns between output images. But the registration result is compromised due to difference in curvature patterns. We suggest a method that demonstrates that there exist some pixels in the output images registered by [2] which have curvature mismatch. Also, this method minimizes the number of pixels having mismatch in curvature patterns between the output images obtained in [2]. The schematic of the proposed method is shown in Fig. 2.

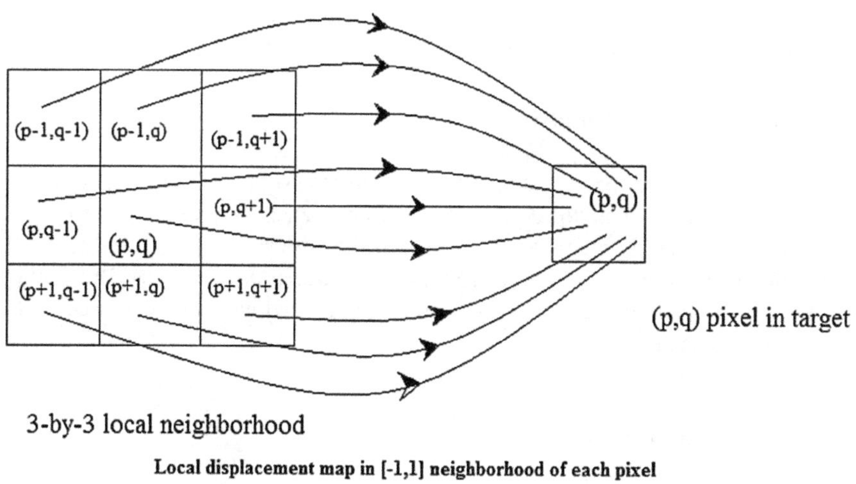

Fig. 2 Illustration of local displacement map in a 3 × 3 window

In this method, firstly, the registered image of the registration method [2] along with source image is taken as input images. This registered image obtained from method [2] is taken as target image for further curvature-based registration process. Mean and Gaussian curvatures of every corresponding pixel in source and target image are obtained as discussed earlier. Then, local surface primitives are checked between corresponding pixels. If the surface primitive do not match, then a local displacement matrix is generated based on the mean and the Gaussian curvature information for every pixel in its neighborhood. Based on this local displacement matrix, every pixel is displaced to its correct position. In this way, the target images are registered with the source image based on the curvature properties. The proposed method is explained next in details.

4.1 *Curvature Extraction of Intensity-Based Registered Image*

Firstly, a source image from brain MRI [11] datasets and the corresponding registered image, obtained from method [2], as the target image are taken as inputs. Then, mean and Gaussian curvature values are obtained for every corresponding pixel in the source and the target images. The method of extraction of mean and Gaussian curvature has already been discussed in Sect. 3. Those curvature values are denoted as (H, K) values where H = Mean curvature and K = Gaussian curvature. Some graphs showing the curvature mismatch have been shown in Sect. 5.1 (refer Figs. 4 and 5).

4.2 *Local Displacement Matrix (LDM)*

4.2.1 Local Surface Primitive

Based on Table 1 and the curvature values obtained, every pixel in target image is labeled according to their local surface primitives. Then, these surface primitives between corresponding pixels are compared if they are same or not. If same, then they belong to same neighborhood and same surface primitive class. But if they are not same, then their curvature difference is compared which is discussed in the succeeding subsection.

4.2.2 Curvature Difference

When the corresponding pixels belong to same local surface primitive class, then a difference metric is used to determine curvature difference between each corresponding pixels in source and target images. The metric is given by:

$$\delta_{\text{curvature}} = \sqrt{\left(H_{\text{source}} - H_{\text{target}}\right)^2 + \left(K_{\text{source}} - K_{\text{target}}\right)^2} \tag{17}$$

where $(H_{\text{source}}, K_{\text{source}})$ and $\left(H_{\text{target}}, K_{\text{target}}\right)$ are corresponding (H, K) values of source and target images, respectively.

4.2.3 Thresholding

The curvature difference metric thus gives us curvature difference values between source and target images. Then, thresholding of those curvatures is done with a chosen threshold value. The pixels below this threshold are the points of interest. Local displacement will have to be generated for those points of interest.

This curvature difference metric between two corresponding pixels is then checked if they are equal or not. If they differ in curvature metric, then a local first-order neighborhood is chosen for each pixel.

4.2.4 Local Displacement Matrix

In the chosen local neighborhood of 8 neighbors, differences in curvature between the pixel in target image and its neighbors are calculated. So, a 3 × 3 matrix containing local curvature differences in the neighborhood is obtained as illustrated in Fig. 3.

In the matrix, the position with minimum curvature difference is found. Now, if two corresponding pixels are found to have minimum curvature difference, then no displacement is required. Otherwise, the target pixel is displaced to the position

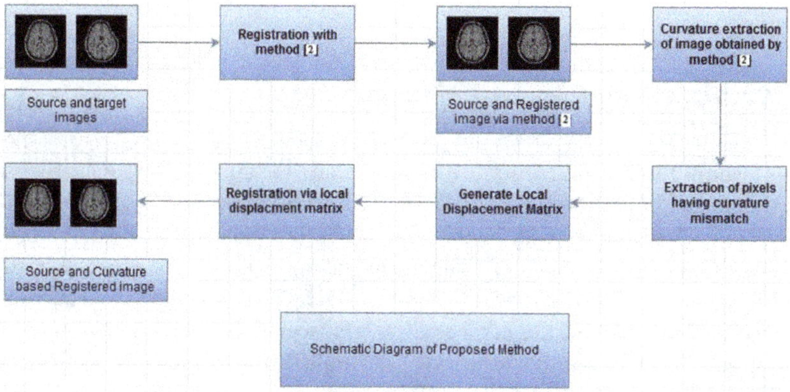

Fig. 3 Schematic diagram of proposed method

with minimum curvature difference. This matrix so generated is called the local displacement matrix. The position where minimum difference is found will then represent the required local displacement of the pixel in the target image. We can observe that more curvature difference for a pixel requires more local displacement, whereas less curvature difference requires less local displacement for that pixel.

In this way, local displacement field for each non-registered pixel is obtained. Then finally, the displacement field should be applied to the floating image pixels to be registered with reference image. Though the method incorporates the curvature, it is based on a two-stage process resulting in higher accuracy at the cost of more computational time. In the next Sect. 5, some tabular data is shown depicting the impact of curvature in registration. The final registration accuracy by application of local displacement map is not shown in this paper as it is still in progress. The curvature property requires very fine refinement of intensity map. So, the 2D space where the displacement may occur should be small compared window size used in intensity-based registration. Hence, one can select a window size, $w \in \{3,4,5,\ldots,31\}$ for LDM.

This method is a two-stage process: firstly, registration using method [2] and then registration using the proposed method. In registration, execution time and registration accuracy are found to be complementary. So, the primary objective of our method is to improve accuracy of the registration may be at the expense of longer time though.

5 Experimental Results

5.1 Curvature Extraction

See Figs. 4 and 5.

5.2 Results Using Proposed Method

The following Table 2 demonstrates the curvature mismatch using the proposed method in terms of number of pixels compared to the method [2]. The number of pixels in registered image from method [2] which are matched with source based on curvature values is calculated. Then after creating local displacement matrix according to the proposed method, the number of matched pixels based on curvature difference is obtained. So, the number of pixels, which is to be aligned based on curvature information, can be obtained. The brain MRI datasets are available at BrainWeb [11–16]. For a typical brain MRI dataset, maximum of 4518 pixels (6.89% of the image pixels) were not registered properly according to intensity in method [2]. On other way, maximum of 2813 pixels (4.29% of image pixels) (refer Table 2) were

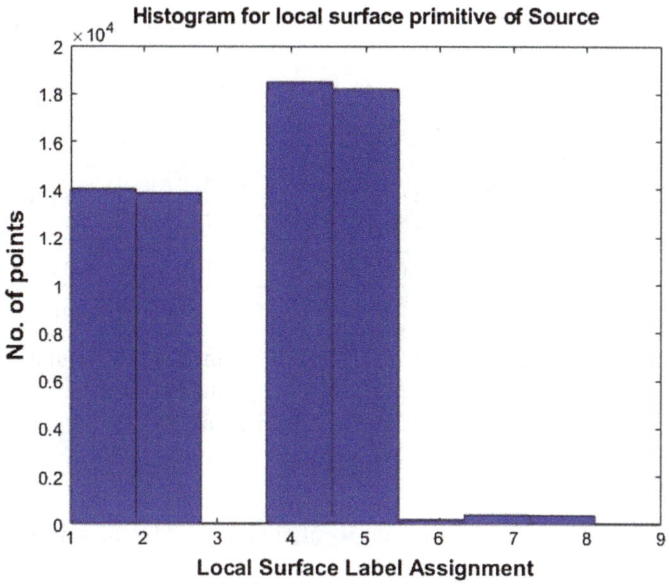

Fig. 4 Brain MRI Dataset 1 [11]: Histogram showing number of pixels assigned surface class labels for source image

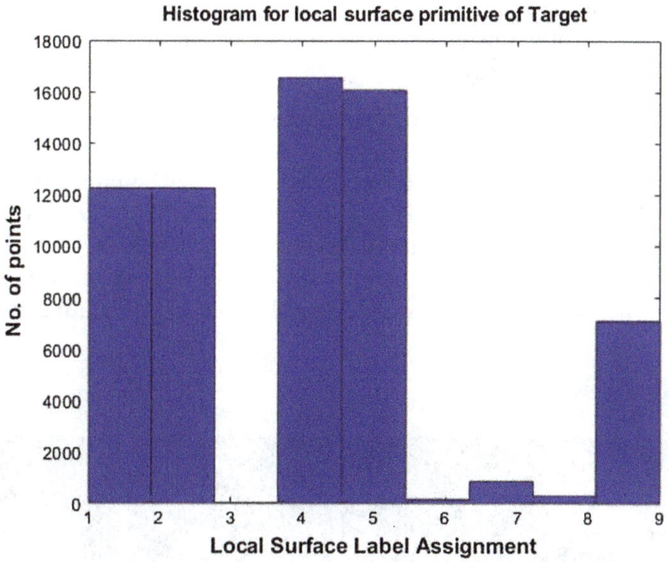

Fig. 5 Brain MRI Dataset 3 [11]: Histogram showing number of pixels assigned surface class labels for target image

Table 2 Comparison between method [2] and proposed method based on number of pixels in brain MRI [11] images

Datasets	N_I	$N_{I,\text{aligned}}$	N_C	$N_{C,\text{aligned}}$
Dataset 1	61018	4518 (6.89%)	62723	2813 (4.29%)
Dataset 2	65529	7 (0.01%)	65536	0
Dataset 3	64592	944 (1.44%)	64824	712 (1.09%)
Dataset 4	65271	265 (0.40%)	65364	172 (0.26%)
Dataset 5	64764	772 (1.18%)	65103	433 (0.66%)

not registered properly according to curvature in our method. This number of pixels can therefore be properly registered based on curvature. However, the remaining pixels (2.6%) were not exactly matched with intensity, but they have same local surface class. This justifies the necessity of curvature, especially in medical image registration.

$\boldsymbol{N_I}$ = No. of pixels matched with intensity after registered by intensity-based registration [2]
$\mathbf{N_{I,aligned}}$ = No. of pixels to be aligned with intensity map in method [2] to improve registration accuracy.
N_C = No. of pixels matched with curvature after creating LDM in the proposed curvature-based method.
$\boldsymbol{N_{C,aligned}}$ = No. of pixels to be aligned with LDM in the proposed method to improve registration accuracy.

Some of the registered outputs of the proposed method are shown in Figs. 6, 7, 8, and 9.

The results have been obtained for our proposed method on a typical desktop with Intel (R) Core (TM) i3-4005U processor with 1.70 GHz, 4.00 GB RAM, and 64-bit Operating System specifications. The registration accuracy, i.e., the quantitative results by applying local displacement map on floating images is not shown in this paper as the work is in progress. Those will be presented in further publication report.

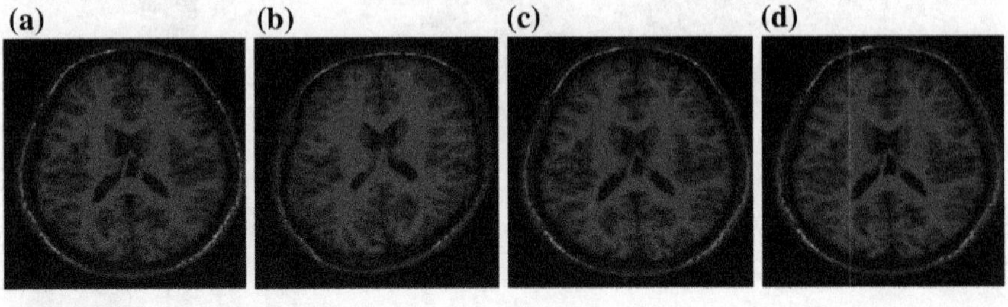

Fig. 6 Brain MRI-Dataset 1 [11]: **a** Source image, **b** Floating image, **c** Registered via [2] **d** Registered via proposed method

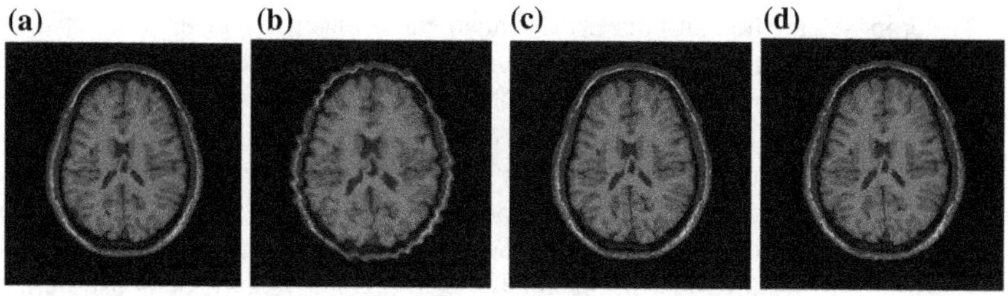

Fig. 7 Brain MRI-Dataset 3 [11]: **a** Source image, **b** Floating image, **c** Registered via [2] **d** Registered via proposed method

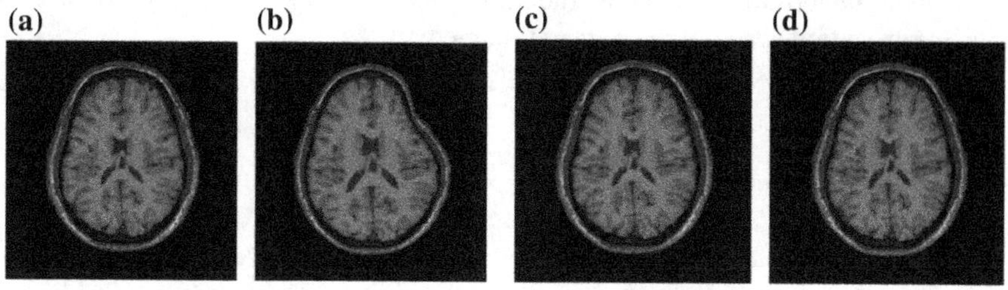

Fig. 8 Brain MRI-Dataset 4 [11]: **a** Source image, **b** Floating image, **c** Registered via [2] **d** Registered via proposed method

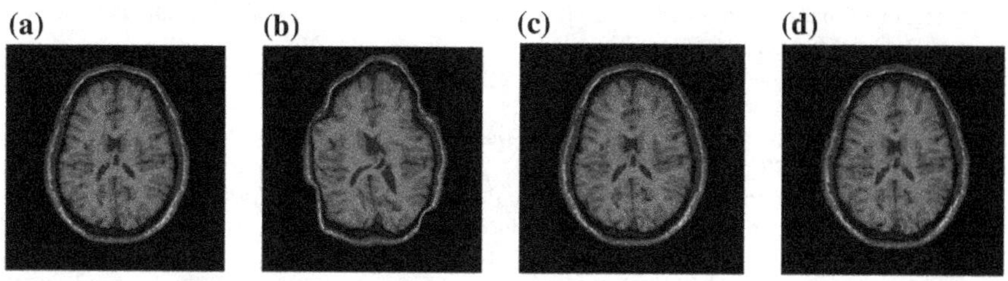

Fig. 9 Brain MRI-Dataset 5 [11]: **a** Source image, **b** Floating image, **c** Registered via [2] **d** Registered via proposed method

6 Conclusions and Future Scope

The problem of non-rigid bio-medical image registration has become, nowadays, a very challenging problem due to its high degrees of freedoms and inherent requirement of smoothness. This paper provides a direction toward improvement of intensity-based non-rigid registration by incorporating curvature information of pixel in an image. From the experimental results and outputs, it is evident that registration can be more accurate if curvature is fused with intensity. Also the outputs of [2] and proposed method are visually same although they have very fine differences which can only be outlined by curvature.

The proposed method introduces the fundamental concepts of mean and Gaussian curvature of a pixel in image and its relation to the non-rigid registration. The method generates a local displacement matrix (LDM) for every pixel in target image for fine tuning of intensity-based registration based on mean and Gaussian curvature information after labeling every pixel according to its local surface primitive class. A Sum of Squared Differences (SSD) metric has been used here for obtaining differences in curvature values between corresponding pixels in source and target images.

Experimental results, both in quantitative and qualitative way, demonstrate the importance of curvature in non-rigid intensity-based image registration.

One future scope is to mathematically model of curvature in order to incorporate inside the energy function required for Graph-Cut-based optimization to bring more automation. Other direction can be the implementation of adaptive window size to enhance execution speed of the registration method.

References

1. Maintz, J.B.A., VandenElsen, P.A., Viergever, M.A.: Evaluation of ridge seeking operators for multimodality medical image matching. IEEE Trans. Pattern Anal. **18**, 353–365 (1996)
2. Chowdhury, S., Roy, R., Bose, S., Khalifa, F., Elnakib, A., El-Baz, A.: Non-rigid biomedical image registration using graph cuts with a novel data term. In: Proceedings Ninth IEEE Int'l Symposium on Biomedical Imaging (ISBI), Barcelona, Spain (2012)
3. Usher, D., Dumskyj, M., Himaga, M., Williamson, T.H., Nussey, S., Boyce, J.: Automated detection of diabetic retinopathy in digital retinal images: a tool for diabetic retinopathy screening. Diabet. Med. **21**, 84–90 (2003)
4. Raussen, M.: Elementary differential geometry: curves and surfaces. Aalborg University, Denmark (2008)
5. Crum, W.R., Hartkens, T., Hill, D.L.G.: Non-rigid image registration: theory and practice. British J. Radiol. **77**, S140–S153 (2004)
6. Boykov, Y., Kolmogorov, V.: An experimental comparison of Min-Cut/ Max-Flow algorithms for energy minimization in vision. IEEE Trans. Pattern Anal. Machine Intell. **26**(9), 1124–1137 (2004)
7. Boykov, Y., Veksler, O., Zabih, R.: Fast approximate energy minimization via graph cuts. IEEE Trans. Pattern Anal. Machine Intell. **23**(11), 1222–1239 (2001)
8. Zhang, W.: Geometry of Curves and Surfaces. Mathematics Institute, University of Warwick, 18 Sep 2014
9. Kalra, P., Peleg, S. (eds.): Computer Vision, Graphics and Image Processing. ICVGIP 2006, LNCS 4338, pp. 228–239. ©Springer-Verlag, Berlin, Heidelberg (2006)
10. Chowdhury, A.S., Burns, J., Sen, B., Mukherjee, A., Yao, J., Summers, R.M.: Detection of pelvic fractures using graph cuts and curvatures. In: Proceedings Eighteenth IEEE Int'l Conference on Image Processing (ICIP), pp. 1605–1608. Brussels, Belgium (2011)
11. Aubert-Broche, B., Evans, A., et al.: A new improved version of the realistic digital brain phantom. NeuroImage **32**(1), 138–145 (2006)
12. http://www.bic.mni.mcgill.ca/brainweb/
13. Cocosco, C.A., Kollokian, V., Kwan, R.K.-S., Evans, A.C.: BrainWeb: Online Interface to a 3D MRI Simulated Brain Database. NeuroImage **5**(4), part 2/4, S425, 1997—Proceedings of 3-rd International Conference on Functional Mapping of the Human Brain, Copenhagen, May 1997
14. Kwan, R.K.-S., Evans, A.C., Pike, G.B.: MRI simulation-based evaluation of image-processing and classification methods. IEEE Trans. Med. Imaging. **18**(11), 1085–1097 (1999)

15. Kwan, R.K.-S., Evans, A.C., Pike, G.B.: An extensible MRI simulator for post-processing evaluation. Visualization in Biomedical Computing (VBC'96). Lecture Notes in Computer Science, vol. 1131, pp. 135–140. Springer (1996)
16. Collins, D.L., Zijdenbos, A.P., Kollokian, V., Sled, J.G., Kabani, J.G., Holmes, C.J., Evans, A.C.: Design and construction of a realistic digital brain phantom. IEEE Trans. Med. Imaging **17**(3), 463–468 (1998)

Prasenjit Kumar Mudi is an assistant professor at Department of Electronics and Communication Engineering, BP Poddar Institute of Management and Technology, India. He received his B.E. and M.E. from Jadavpur University, India, in 2012 and 2016. His research interests are Computer Vision, Image Processing with focus on medical applications.

Signal Processing and Nuclear Structure of ^{106}Cd Nucleus

D. Choudhury and R. Goswami

Abstract The study of the structure of nucleus ^{106}Cd has been undertaken using conventional modern accelerators, state-of-the-art detectors coupled to high-end NIM compatible electronic modules, and a high end data acquisition system for the analysis of the raw data acquired in list mode. In the present work, in-beam gamma-ray spectroscopy was performed via fusion evaporation reaction and the structure of the nucleus from the high-spin state down to the ground state was studied systematically. Two new gamma-ray transitions from known energy levels have been reported. The sequence of E1 transitions depopulating alternating parity levels and the new transition of octupole multipolarity signifies possible octupole correlation hitherto unreported for nuclei in this mass region.

Keywords Accelerators · Signal processing · Nuclear structure · Octupole correlations

1 Introduction

The study of nuclear structure of various isotopes in the periodic table is done following the nuclear techniques involving state-of-the-art ion sources, accelerators, modern detectors, associated electronics set up and a correlated data acquisition system [1]. One of the methods of investigating nuclear structure is in-beam gamma-ray spectroscopy using fusion evaporation reaction. Light or heavy ion beams of medium energy (1 ~ 2 MeV/nucleon) from ion sources are accelerated through cyclotron, pelletron, linac accelerators and are allowed to collide and fuse with the target nucleus in the target chamber to form a compound nuclear system [2]. The incoming beam nuclei have enough energy to overcome the Coulomb barrier between the target and projectile prior to the formation of the compound nuclear system. This is a hot nuclear system which attains a thermodynamic equilibrium within 10^{-20} s after which it cools by emitting nucleons, alphas and gamma rays.

D. Choudhury · R. Goswami (✉)
B.P.Poddar Institute of Management and Technology, 137, VIP Road, Kolkata 700052, India
e-mail: hodbasicsc1@gmail.com

J. K. Mandal et al. (eds.), *Information, Photonics and Communication*,
Lecture Notes in Networks and Systems 79,
https://doi.org/10.1007/978-981-32-9453-0_22

The total angular momentum is conserved in this process. Hence, the compound nucleus is formed at very high excitation energy depending on the angular momentum carried by the accelerated projectile. While cooling from the excited state, the compound nucleus initially evaporates particles until the system reaches a state where the final excitation energy of the concerned state is less than the particle separation energy above the *yrast* line. The daughter nucleus now cools to lower excitation energy by emitting γ-rays until the nucleus reaches the ground state. The structure of the nuclei is studied while the daughter nucleus de-excites to the ground state with the emission of γ-rays. The de-exciting γ-rays act as an input signal to the germanium clover detectors surrounding the target chamber where the compound nuclear reaction has taken place.

The γ-rays emitted are captured by the detector arrays [3] surrounding the target nucleus. The germanium clover detectors serving as gamma-ray arrays are coupled to Compton-suppressed spectrometer (CSS). This anti-Compton shield (CSS) is used to reduce the Compton background around a photopeak, thereby improving the signal to noise ratio. In our work, we have used the INGA (Indian National Gamma Array)

Figure 1 comprises six clover detectors. These radiation detectors operate in pulse mode the output of which gets converted to linear pulse whose amplitude and shape have all the information of interest. To study particular nuclei of interest excitation function, measurements are performed where the optimum beam energy is obtained to maximize the cross section and angular momentum input for the channel of interest. In recent years, the use of Indian National Gamma Array (INGA) has opened the possibility to perform high-resolution discrete gamma-ray spectroscopy of excited

Fig. 1 Picture of Indian National Gamma Array at TIFR-BARC Pelletron linac facility [3]

nuclei produced through various reactions where shape coexistence, octupole correlation, magnetic/anti-magnetic rotation, chiral bands, high-spin states and many other characterizations of the nuclear structure has been made possible. This detector array is designed for twenty-four Compton-suppressed clover detectors arranged in a spherical geometry (Fig. 1) with six detectors at 90° and three detectors each at 23°,40°,65°,115°,140° and 157° with respect to the beam direction. With the clover crystals set at 25 cm from the target, the overall photopeak efficiency is around 5% at Eγ ~ 1 MeV. Four n-type crystals in a single cryostat form a single-clover detector. Each crystal has a preamplifier with a gain of 200 mV/MeV and decay time constant 50 μs. The BGO shield used for Compton suppression has sixteen photomultipliers the sum of which is sent to a timing filter amplifier and then to the analog constant fraction discriminator to generate the NIM logic signal. The signal is the input to the digital data acquisition system used for the acquisition of the raw data in list mode. The offline analysis of this raw data using state-of-the-art software generates a 4096 × 4096 matrix for further analysis and nuclear structure study.

2 The Experiment

Following the detailed signal processing described above, the excited states of ^{106}Cd nucleus were populated in the ^{93}Nb(^{20}Ne, 3p4n) ^{106}Cd at a beam energy of E = 150 MeV at the Variable Energy Cyclotron Centre, Kolkata. The experiment yielded about ~10^9 two or higher-fold coincident events obtained with eight Compton-suppressed clover detectors of INGA the detailed discussion of which is given in [3]. In the present INGA setup at VECC, we have used four clover detectors at 90°, two each at 125° and 40° to the beam direction. Gated spectra with a dispersion of 0.5 keV per channel were generated using the software INGASORT from a large dimension of 4096 × 4096 matrices. These large data matrices were obtained from the sorting of the gain matched raw data. The symmetric Eγ–Eγ matrix was used to establish the level scheme and study nuclear structure through γ-ray coincidence relationships and intensity arguments. The DCO (directional correlation of oriented nuclei) ratio determines the multipolarity of the gamma-rays de-exciting from various nuclear levels. This was obtained by sorting the asymmetric matrix with events recorded by the clover detectors set at 90° along the y-axis and remaining clover detectors at other angles along the x-axis [4].

3 Experimental Result

In the present study of the nuclear structure of ^{106}Cd, a search for octupole correlations has been initiated. A new structure Fig. 2 of ^{106}Cd at medium spin has been reported in the present study. Excepting for ^{108}Cd, this structure was not reported for any nuclei in this mass region. The current study also proposes the possibility of observing similar

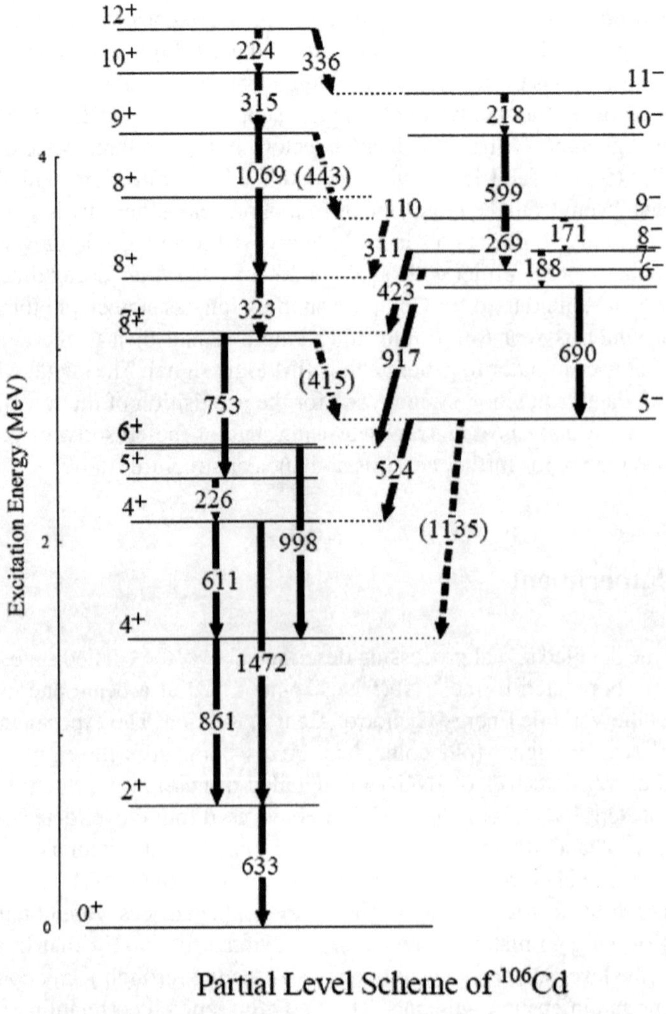

Fig. 2 Partial level scheme of ^{106}Cd showing the relevant transitions

characteristic for nuclei in this mass region. Such structures have been reported in the A ~ 120 mass region [5], where octupole correlations have been reported down to low-lying even parity negative spin levels. Octupole correlations in nuclear structure come with alternate parity sequence of levels linked by enhanced E1 transition and collective E3 transitions and observation of parity doublets occurring due to the proximity of orbitals of opposite parity resulting in alternating parity sequence of levels are some signatures of the reported octupole correlations.

In the present work, we observe alternate parity sequence of levels connected by enhanced E1 transitions. Gamma transitions of 415, 1135, 443 keV of E3 and E1 multipolarity have been observed for the first time in the present work along with

335, 463 and 524 keV E1 transition observed from the previously reported [4] data. A relevant gate set on 415 keV gamma ray has been shown in Fig. 3. This transition has been conjectured as a possible E3 transition because it occurs between states having $\Delta J = 3$ with a change in parity between the two states. The DCO ratio measurements from the present work Fig. 4 confirm the $\Delta J = 3$ multipolarity of the newly observed transition which also signifies the onset of octupole correlations.

The experimental DCO ratio data for the present reaction is shown in Fig. 4.

The partial level scheme of Fig. 2 shows the new transitions with dotted lines. The reduced transition probability obtained from the previously reported data for 335 keV $E1$ transition is 2.7×10^{-5} e^2 fm^2, and the same for the 415 keV $E3$ transition is 2.3×10^5 e^2 fm^6. These transition probabilities are relatively enhanced compared to their single-particle estimates. The signature of octupole correlation is also studied from the observation of the relative placement of the positive and negative parity levels. A negative parity level of spin I^- for a nucleus bearing octupole correlation is lower in energy than the $(I + 1)^+$ levels. The initial yrast negative parity states in the present work are higher in energy than the positive parity states. But at higher excitation inversion occurs showing the slow onset of octupole correlations. Such onset of quadrupole–octupole correlations has already been reported in the neighbouring ^{108}Cd nucleus [6]. Hence, we observe in Fig. 2 that the 7^- spin state has lower excitation energy [3.4 MeV] than the corresponding 8^+ spin state [3.8 MeV].

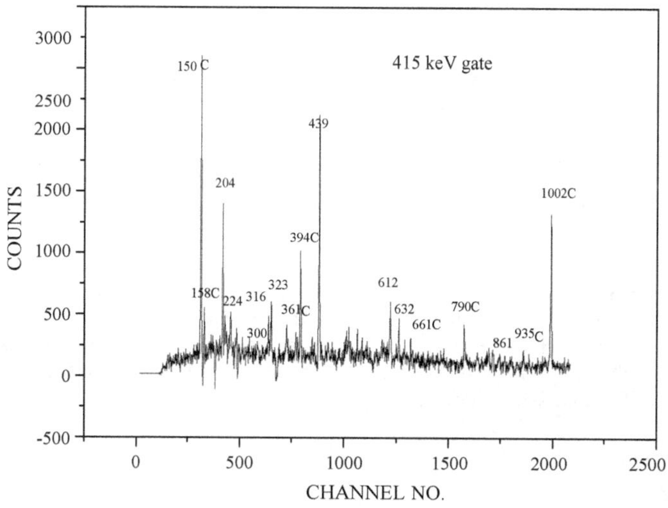

Fig. 3 Spectrum obtained from the 415 keV gate

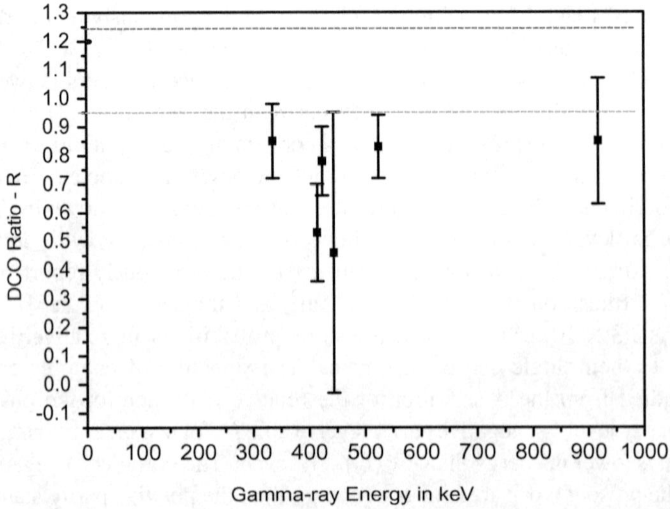

Fig. 4 Experimental DCO ratios for dipole and octupole transitions reported previously and also observed in the present work are shown in the figure. The line in green and red corresponds to expected values of the stretched dipole and quadrupole transitions, respectively

4 Conclusion

The present work shows the onset of octupole correlations in ^{106}Cd nucleus. At higher excitations, all negative parity levels have lower excitation energy than the corresponding positive parity states. The single-particle transition probability shows enhanced E1 and E3 transitions. The DCO ratio measurements provide conclusive evidence about the dipole multipolarity of the γ-transition observed in the present work.

Acknowledgements The authors gratefully acknowledge the support received from Dr. S. S. Ghugre and Dr. R. Raut at every stage of this work. The help and support received from the INGA collaboration in setting up the array at VECC, and support from the Cyclotron staff is gratefully acknowledged. They also wish to acknowledge the grant received from UGC-DAE-CSR-KC under the CRS: UGC-DAE-CSR-KC/CRS/13/NP07.

References

1. Glenn, K.: Radiation Detection and Measurement, 3rd edn. Wiley Inc, New York (1999)
2. Regan, P.: Post Graduate Nuclear Experimental Techniques Course Notes (2003)
3. Palit, R. et al.: Performance of Indian National Gamma Array (INGA) coupled with a fast digital data acquisition system for nuclear structures studies. J. Phys. Conf. Ser. 420012159
4. Regan, P.H., Stuchbery, A.E., Dracoulis, G.D., et al.: High spin proton and neutron intruder configuration 106Cd. Nucl. Phys. A **586**, 351–376 (1995)

5. Mason, P., Benzoni, G., Bracco, A., et al.: Evidence of Octupole correlations in ^{124,125}Ba. PRC **72**, 064315 (2005)
6. Gade, A., Belie, D., von Brentano, P., et al.: Dipole excitations ^{108}Cd. PRC **67**, 034304 (2003)

Performance Study of Some Recent Optimization Techniques for Energy Minimization in Surveillance Video Synopsis Framework

Subhankar Ghatak and Suvendu Rup

Abstract In the age of the smart city, each activity is under surveillance. The employment of plentiful surveillance video cameras produces the gigantic amount of redundant video data. For ease of investigations, video synopsis competently shrinks the length with the preservation of all activities presents in the original video. The outcome of the video synopsis technology greatly depends on the central module, the optimization framework, and its minimization. This paper evaluates the performance of various optimization techniques, namely simulated annealing (SA), NSGA II, cultural algorithm (CA), teaching–learning-based optimization (TLBO), gray wolf optimizer (GWO), forest optimization algorithm (FOA), JAYA algorithm, elitist-JAYA algorithm, self-adaptive multi-population-based JAYA algorithm (SAMP-JAYA), to minimize the energy in the field of object-based surveillance video synopsis. The experimental results and analysis direct the need for an optimization algorithm which can efficiently and consistently solve the minimization problem in connection to video synopsis.

Keywords Optimization · Energy minimization · Video synopsis · Video surveillance

1 Introduction

In the age of big data, a huge amount of surveillance video data is captured and transmitted daily, through the working of a numerous number of surveillance cameras. By nature, the surveillance video data are of long duration and redundant. Effective inspection of these videos demands an efficient technology which reduces the

S. Ghatak (✉) · S. Rup
Image and Video Processing Laboratory, Department of Computer Science and Engineering, International Institute of Information Technology, Bhubaneswar 751003, India
e-mail: subhankar@iiit-bh.ac.in

S. Rup
e-mail: suvendu@iiit-bh.ac.in

J. K. Mandal et al. (eds.), *Information, Photonics and Communication*, Lecture Notes in Networks and Systems 79,
https://doi.org/10.1007/978-981-32-9453-0_23

labor of watching these boring long-duration videos by shortening the length with a preservation of all activities present in the original one. Video synopsis (VS) is such a prominent technology. VS effectively reduces the length with successful preservation of all activities. In terms of investigation and browsing, synopsis video is user-friendly.

The authors in [1] pioneered VS technology and the concept of object-based video synopsis is introduced. The optimization framework is considered to be the main building block of object-based video synopsis generation among the other preprocessing modules (i.e., object detection and segmentation and tracking) and post-processing module (i.e., stitching). From the inception to till date, several researches [2, 3] and [4], etc., are carried out in this field. Majority focused on the generation of video synopsis and rest concentrate on the quality enhancement factor in terms of visualization of the final synopsis.

In this work, a performance evaluation study is carried out among various optimization algorithms, namely SA [5], NSGAII [6], CA [7], TLBO [8], GWO [9], FOA [10], JAYA [11], Elitist-JAYA [12], and SAMP-JAYA [13] to minimize the energy of the original video. The major contribution of this work is to direct the effective use and selection of the proper optimization algorithm for energy minimization in object-based surveillance video synopsis generation.

The rest of the paper is organized as follows. Section 2 presents the formulated surveillance video synopsis framework in details. Experimental results and analysis are presented in Sect. 3. Finally, the conclusive discussion is outlined in Sect. 4.

2 Surveillance Video Synopsis Framework

A single camera with a static background and object-based surveillance video synopsis framework constitutes of three major phases. First, the preprocessing phase, which includes object detection and segmentation module followed by a multi-object tracking process. The second phase is the central phase of the video synopsis generation, which is the optimization framework along with energy minimization. The last phase contains the stitching module to obtain the final video synopsis. This section explains the different phases of video synopsis generation in details.

2.1 Preprocessing: Object Detection & Segmentation and Tracking

This is the initial step of an object-based video synopsis generation. In an object-based video synopsis philosophy, not all the moving objects should draw the users attention & not all static objects can be ignored. For example, the shaking of leaves is not the significant information, on the other hand, a person with no motion activity may be considered as an important activity in the video synopsis problem. So taking the aforementioned fact into account and similar with [1–3] and [4], in this work, a background subtraction method is employed for object detection segmentation. In case of a static surveillance camera, the background usually changed due to the varying illumination or the camera vibration. Thus, the background is formed through a simulation of temporal median over a period of 30 s (15 s before and after of the current frame). Then, the corresponding background frame is subtracted from the current frame to form a foreground mask. Two-dimensional morphological erosion and dilation are applied to the obtained foreground mask for the generation of more accurate result. Blob analysis is applied then on the foreground for detection. The detected results are tracked through the widely used multi-object tracker Kalman Filter [14], and the object tubes are formed as connected bounding boxes in the time axis. The resulted objects tubes are passed to the next phase for further processing.

2.2 Optimization Framework and Energy Minimization

To obtain the optimized length of the synopsis video, which includes all objects' activities, an energy minimization optimization framework in terms of activity cost, E_a, collision cost, E_c, and temporal consistency cost, E_t, is formulated as in Eq. (1).

$$E(f_\mu) = E_a(f_\mu) + \omega_0 E_c(f_\mu) + \omega_1 E_t(f_\mu) \tag{1}$$

where f_μ is a mapping of objects from input video to the synopsis video and the weights ω_0 and ω_1 are importance of the corresponding costs, given by the user.

$E_a(f_\mu)$, the penalty for losing any activity from the original video and is described in Eq. (2).

$$E_a(f_\mu) = \sum_{o \in O} E_a(o^s) = \sum_{o \in O} \sum_{\hat{o} \in o^s \wedge \hat{o} \notin synopsis} absDiff(\hat{o}) \tag{2}$$

Here O is a set of all the object activity tubes present in the original video, o is a tube element that is in O and o^s is a mapping result of o. Let, $\hat{o}$ be an instance element that is in O but not included in the synopsis. The absolute difference between $\hat{o}$ and its corresponding background is formulated as $absDiff()$ and is given in Eq. (3).

$$absDiff(\hat{o}) = \sum_{(x,y) \in bbox(\hat{o})} ||V(x, y, t_{\hat{o}}) - B(x, y, t_{\hat{o}})|| \tag{3}$$

Here V and B denote the original video frames and the corresponding background model, respectively, and $bbox()$ stands for bounding box function. Activity cost becomes zero if all activities are well preserved in the synopsis video (i.e. lossless synopsis).

As the synopsis video has a smaller time period as compared to the original video, the presence of collision becomes obvious and the penalty imposed for this occurrence is denoted by $E_c(f_\mu)$. In Eq. (4), o_m^s & o_n^s are the mapping objects to the synopsis video from the original video elements, o_m & o_n. Reducing the collision between objects would surely enhance the synopsis quality.

$$E_c(f_\mu) = \sum_{o_m, o_n \in O} E_c(o_m^s, o_n^s) = \sum_{o_m, o_n \in O} \sum_{\hat{o}_m \in o_m^s, \hat{o}_n \in o_n^s} Area(bbox(\hat{o}_m) \cap bbox(\hat{o}_n)) \tag{4}$$

Here $Area()$ function evaluates the overlap between the bounding boxes of the objects and the cost becomes zero if no collisions are detected in the produced synopsis video.

The process for the generation of a synopsis video may cause irregularities in the temporal connotations among objects owing to the variation in time of appearance. The penalty allotted for the irregularities among objects with respect to the original video is denoted by $E_t(f_\mu)$, and it aims at reducing the deviance from the original order. Two different situations are exhibited to explain the process for assigning of the penalty. Firstly, in the case where two objects share common frames and in other situations, two objects do not share any common frames. Further, the second situation is classified into two sub-categories with respect to violation or preservation of chronology and entry difference of objects of synopsis and the original video.

Spatial relationships (Δ) between two moving objects can be utilized to measure their probability of interaction. This measurement represents the first considered situation. The spatial relationships between two moving objects, say o_m and o_n, is established as:

$$\Delta(o_m, o_n) = \exp(-min_{t \in t_{om} \cap t_{on}}(\delta(o_m, o_n, t))/\sigma) \tag{5}$$

Here the Euclidean distance between the objects o_m and o_n from the original video in t^{th} frame is represented as $\delta(o_m, o_n, t)$, where σ is used to tune the level of space communication between o_m and o_n.

Likewise, the temporal relationships (τ) between objects can be expressed as:

$$\tau(o_m, o_n) = ||(\vec{t}_{o_m} - \vec{t}_{o_n}) - (\vec{t}_{o_m^s} - \vec{t}_{o_n^s})|| \tag{6}$$

Here $\vec{t}_{o_m^s}$ and $\vec{t}_{o_n^s}$ are the entry frame indices in the synopsis video and $\vec{t}_{o_m}$ and $\vec{t}_{o_n}$ are that of o_m and o_n, respectively, in the original video.

The measurement of the temporal violation ($\hat{\tau}$) between o_m and o_n is defined as:

$$\hat{\tau}(o_m, o_n) = (\vec{t}_{o_m} - \vec{t}_{o_n}) \times (\vec{t}_{o_m^s} - \vec{t}_{o_n^s}) \tag{7}$$

The positive value of $\hat{\tau}$ implies successful preservation of temporal consistency between o_m and o_n in the synopsis incurring zero penalty cost. Thus, based on the considered situations, the complete representation of the temporal consistency cost is represented as:

$$E_t(f_\mu) = \sum_{o_m, o_n \in O} E_t(o_m^s, o_n^s) \tag{8}$$

$$E_t(o_m^s, o_n^s) = \begin{cases} \triangle(o_m, o_n) \times \tau(o_m, o_n) & \text{if } (t_{o_m} \cap t_{o_n}) \neq \phi \\ \begin{cases} \exp(\|(\vec{t}_{o_m^s} - \vec{t}_{o_n^s})\|/\gamma) & \text{if } \hat{\tau}(o_m, o_n) \leq 0 \\ 0 & \text{Otherwise} \end{cases} & \text{Otherwise} \end{cases} \tag{9}$$

In Eq. (9), the first statement is responsible to assign penalty for the situation where two objects o_m and o_n with time durations t_{o_m} and t_{o_n}, respectively, having interaction originally. Otherwise, if the temporal consistency is violated, an exponential cost is assigned where γ is used as a normalization factor. In the third statement, zero cost will be assigned as no temporal consistency is violated.

Minimization of the energy function in Eq. (1) is very exhaustive because of the fact that it searches through an enormous amount of possibilities. From literature, it is evident that SA is widely employed to solve the objective function stated in Eq. (1). This work investigates the performance of various recent optimization algorithms such as NSGA II, CA, TLBO, GWO, FOA, JAYA, Elitist-JAYA, and SAMP-JAYA, for solving the same.

2.3 Stitching

According to the mapping results obtained from the previous module, the object tubes and their corresponding time stamp are stitched together with the synopsis background frames using Poisson image editing [15]. Figure 1 shows the final synopsis video frames.

Fig. 1 Final synopsis video frames

3 Experimental Results and Analysis

3.1 Experimental Setup and Video Data Set

To study the performance of the considered optimization techniques in minimizing the energy cost as defined in Eq. (1), numerous experiments are carried out. For this study, MATLAB (version 2018a) with Intel Core I7 3.2 GHz CPU with 32 RAM is employed. The complete view of the experimental setup is depicted in Fig. 2.

For the validation of this performance study, various videos with different frame rates, video length, and a number of objects are considered. A snapshot of the first frame and different features of the considered videos is presented in Table 1. The videos which are considered for this study have different types of objects. The 1st video, which is taken from [16] contains humans walking and the 2nd video, taken from [17], which was captured within a park involves walking as well as cycling action by humans. Whereas the 3rd video is taken from [18], a scene from MIT campus with a group of people moving around, it also contains tree branches and car movement. The 4th one is a real-generated video taken from the IIIT Bhubaneswar surveillance dataset.

3.2 Performance Evolution of Optimization Framework

The considered optimization algorithms, SA, NSGAII, CA, TLBO, GWO, FOA, JAYA, Elitist-JAYA, and SAMP-JAYA are employed for the energy minimization (given by Eq. (1)), which produce the optimized video synopsis. The corresponding values of the biasing coefficients ω_0 and ω_1 are considered to be 0.8 and 0.2, respectively, overbroad experimentation. Collision cost is assigned a higher bias as per the sheer choice of the user to eliminate collision which causes loss of object activity data.

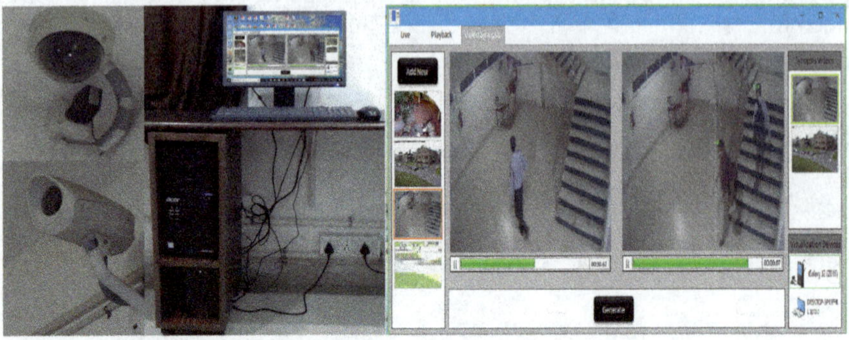

Fig. 2 Experimental setup

Table 1 Parameters of experimental surveillance videos

Video number	Video length (# Frames)	Frame rate (fps)	Number of objects	First frame snap shot
1	600	30	6	
2	1066	30	5	
3	2688	30	12	
4	3810	10	30	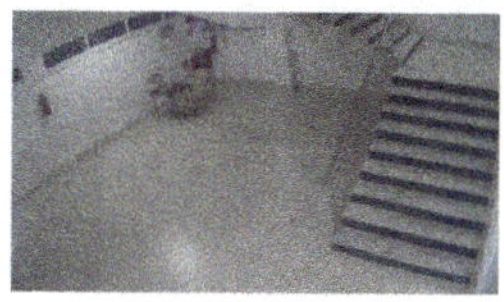

A relative experimental study is also executed among various optimization techniques (i.e., SA, NSGAII, CA, TLBO, GWO, FOA, JAYA, Elitist-JAYA, and SAMP-JAYA) to perceive the potential and precision for the optimization of Eq. (1), and the results are presented in Table 2. The common parameters like the population size and a maximum number of cost function evaluation are chosen as 10 and 100, respectively. In this study, the other algorithm-specific parameter values are considered to be same as specified in the corresponding literature. In this table, the individual costs and the corresponding fitness values along with the execution time are compiled. The length of the synopsis produced is assigned to be equal to that of the longest object tube; thus, a zero activity cost is obtained.

A statistical analysis in terms of best, mean, worst, standard deviation, and average execution time is presented in Table 3 for all the optimization techniques applied to videos 1, 2, 3, and 4, respectively. This analysis is performed using ten independent runs for each algorithm, and the best convergence plot is represented in Fig. 3. In summary, from Tables 2 and 3, it is evident that the existing algorithms are not able to consistently minimize the objective function, defined in Eq. (1), for all considered

Table 2 Performance comparison among various optimization techniques

	Optimization techniques	Activity cost	Collision cost ($\times 10^3$)	Temporal consistency cost	Fitness value ($\times 10^3$)	Time of execution (s)
Video-1	SA	0	00013.77	009.69	00011.19	0207.18
	NSGA II	0	00013.21	**008.38**	00010.73	0047.09
	CA	0	**00008.16**	008.84	**00006.76**	**0022.79**
	TLBO	0	00013.21	008.40	00010.73	0094.13
	GWO	0	00013.44	008.63	00010.92	0074.90
	FOA	0	00015.36	009.18	00012.47	0828.29
	JAYA	0	00013.21	008.40	00010.73	0041.28
	Elitist-JAYA	0	00013.15	009.09	00010.69	0080.65
	SAMP-JAYA	0	00013.10	008.95	00010.65	0081.52
Video-2	SA	0	**00212.85**	009.77	**00172.90**	0120.39
	NSGA II	0	00520.75	011.53	00423.01	0033.12
	CA	0	00271.30	**008.63**	00220.38	**0012.42**
	TLBO	0	00520.75	011.53	00423.01	0053.23
	GWO	0	00233.43	010.90	00189.62	0046.54
	FOA	0	00557.51	011.22	00452.87	0302.48
	JAYA	0	00520.75	011.53	00423.01	0026.89
	Elitist-JAYA	0	00520.80	010.95	00423.05	0052.73
	SAMP-JAYA	0	00520.77	010.87	00423.02	0054.28
Video-3	SA	0	**00005.01**	036.75	**00004.08**	1213.30
	NSGA II	0	00272.19	041.29	00221.10	0266.65
	CA	0	00036.24	041.16	00029.44	**0101.17**
	TLBO	0	00276.72	032.51	00224.78	0512.52
	GWO	0	00015.81	028.39	00012.85	0397.49
	FOA	0	00418.21	**017.58**	00339.48	1268.55
	JAYA	0	00162.33	048.37	00131.87	0269.04
	Elitist-JAYA	0	00006.77	040.69	00005.51	0538.28
	SAMP-JAYA	0	00010.34	038.43	00008.41	0540.95
Video-4	SA	0	14965.34	294.99	12156.40	1015.00
	NSGA II	0	76742.08	308.15	62337.65	0247.75
	CA	0	58961.68	366.92	47895.00	0212.45
	TLBO	0	80458.35	**256.35**	65356.36	0442.28
	GWO	0	43681.65	312.48	35482.67	0280.53
	FOA	0	89242.08	490.15	72483.00	1768.83
	JAYA	0	07770.55	322.65	61885.00	**0149.50**
	Elitist-JAYA	0	35172.00	312.48	28570.00	0603.29
	SAMP-JAYA	0	**07586.81**	334.71	**06162.77**	0661.23

videos. Thus, there is a prominent need for research to evolve an efficient optimization technique for solving the objective function in connection to object-based surveillance video synopsis.

Table 3 Comparative statistical analysis of various optimization techniques

	Statistical measures					
	Optimization techniques	Best ($\times 10^3$)	Mean ($\times 10^3$)	Worst($\times 10^3$)	Standard deviation ($\times 10^3$)	Average execution time (s)
Video-1	SA	00011.19	00014.30	00019.39	0002.86	0202.69
	NSGA II	00010.73	00015.02	00020.94	0004.33	0045.63
	CA	**00006.76**	**00011.98**	00017.12	0003.71	**0022.50**
	TLBO	00010.73	00012.48	**00014.40**	**0001.53**	0096.52
	GWO	00010.92	00014.45	00019.50	0005.69	0081.29
	FOA	00012.47	00016.07	00025.25	0005.23	0593.03
	JAYA	00010.73	00014.30	00020.43	0003.57	0044.68
	Elitist-JAYA	00010.69	00013.86	00018.95	0003.32	0092.18
	SAMP-JAYA	00010.65	00013.77	00017.43	0003.24	0094.54
Video-2	SA	**00172.90**	**00276.80**	**00364.36**	0054.02	0117.81
	NSGA II	00423.01	00454.01	00583.30	0041.11	0032.96
	CA	00220.38	00338.00	00456.70	0111.37	**0013.63**
	TLBO	00423.01	00424.83	00440.89	**0005.64**	0051.94
	GWO	00189.62	00423.82	00435.38	0029.84	0047.49
	FOA	00452.87	00469.86	00487.71	0014.20	0228.29
	JAYA	00423.01	00426.61	00442.68	0007.64	0026.55
	Elitist-JAYA	00423.05	00430.07	00438.69	0007.25	0054.45
	SAMP-JAYA	00423.02	00428.03	00436.75	0007.12	0053.78
Video-3	SA	**00004.08**	**00009.18**	**00018.78**	**0005.45**	1150.33
	NSGA II	00221.10	00303.37	00454.78	0069.52	0270.13
	CA	00029.44	00121.61	00250.80	0080.39	**0105.65**
	TLBO	00224.78	00298.06	00341.17	0037.12	0508.85
	GWO	00012.85	00015.71	0019.58	0021.31	0372.59
	FOA	00339.48	00569.37	00799.93	0066.82	1267.43
	JAYA	00131.87	00226.47	00389.91	0089.27	0259.01
	Elitist-JAYA	00005.51	00018.47	00023.76	0011.54	0503.81
	SAMP-JAYA	00008.41	00025.87	00052.65	0032.59	0509.32
Video-4	SA	12156.40	17750.41	22076.02	2886.88	1033.79
	NSGA II	62337.65	68689.44	72483.18	2913.78	0250.90
	CA	47895.00	56131.40	66194.00	5197.50	0233.99
	TLBO	65356.36	68308.66	69779.14	1223.55	0451.72
	GWO	35482.67	36219.17	41296.73	1594.25	0297.17
	FOA	72483.00	79103.50	85724.00	2518.58	1950.40
	JAYA	61885.00	66565.01	69183.41	2315.49	**0114.43**
	Elitist-JAYA	28570.00	36466.50	44363.00	**0206.54**	0903.82
	SAMP-JAYA	**06162.77**	**06487.64**	**06923.66**	0235.23	0929.89

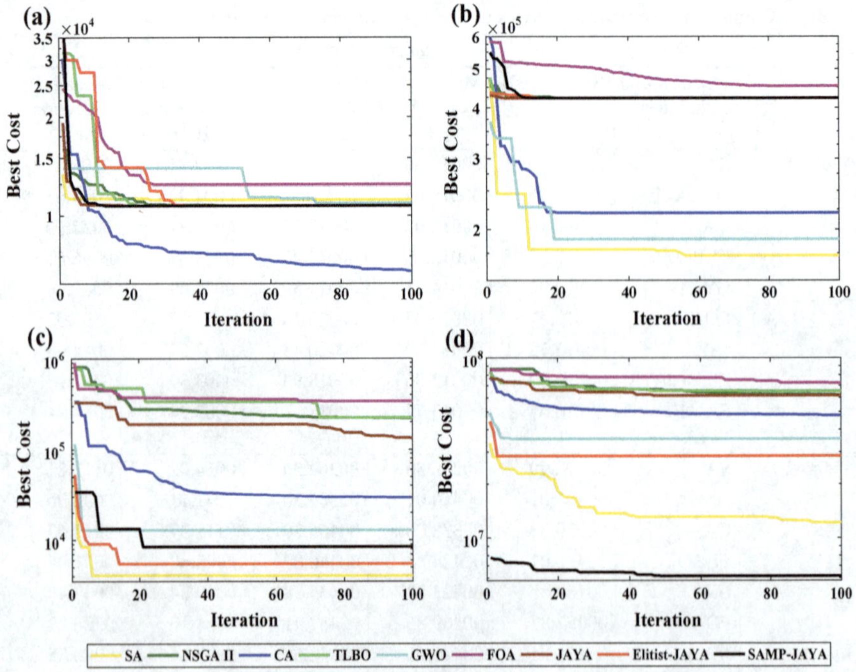

Fig. 3 Comparison of convergence characteristics for various optimization techniques (SA, NSGAII, CA, TLBO, GWO, FOA, JAYA, Elitist-JAYA, and SAMP-JAYA) applied to **a** Video-1 **b** Video-2 **c** Video-3 **d** Video-4

4 Conclusion

Video synopsis has competently proven itself in summarizing long-duration surveillance videos for easy and quick browsing and investigations. The central module, optimization framework, and energy minimization are controlling the outcome, i.e., the synopsis video. From this study, it is observed that the existing meta-heuristic techniques are unable to minimize the energy consistently, though they are widely employed in minimizing the energy for video synopsis generation. Thus, there is a crisis need of an efficient and consistent optimization algorithm, which specifically addresses the said problem.

References

1. Rav-Acha, A., Pritch, Y., Peleg, S.: Making a long video short: Dynamic video synopsis. In: Proceeding CVPR, vol. 1, pp. 435–441. New York, NY, USA (2006)
2. Pritch, Y., Rav-Acha, A., Peleg, S.: Nonchronological video synopsis and indexing. IEEE transactions on pattern analysis and machine intelligence **30**(11), 1971–1984 (2008)

3. Nie, Y., Xiao, C., Sun, H., Li, P.: Compact video synopsis via global spatiotemporal optimization. IEEE Trans. Vis. Comput. Graph. **19**(10), 1664–1676 (2013)
4. Li, X., Wang, Z., Lu, X.: Surveillance video synopsis via scaling down objects. IEEE Trans. Image Process. **25**(2), 740–755 (2016)
5. Kirkpatrick, S., Gelatt, C.D., Vecchi, M.P.: Optimization by simulated annealing. Science **220**(4598), 671–680 (1983)
6. Deb, K., Pratap, A., Agarwal, S., Meyarivan, T.: A fast and elitist multiobjective genetic algorithm: NSGA-II. IEEE Trans. Evol. Comput. **6**(2), 182–197 (2002)
7. Jin, X., Reynolds, R.G.: Using knowledge-based evolutionary computation to solve nonlinear constraint optimization problems: a cultural algorithm approach. In: Evolutionary Computation, 1999. CEC 99. Proceedings of the 1999 Congress on. vol. 3, pp. 1672–1678. IEEE (1999)
8. Rao, R.V., Savsani, V.J., Vakharia, D.: Teaching-learning-based optimization: a novel method for constrained mechanical design optimization problems. Comput. Aided Des. **43**(3), 303–315 (2011)
9. Mirjalili, S., Mirjalili, S.M., Lewis, A.: Grey wolf optimizer. Advances in Eng. Software **69**, 46–61 (2014)
10. Ghaemi, M., Feizi-Derakhshi, M.R.: Forest optimization algorithm. Expert Syst. Appl. **41**(15), 6676–6687 (2014)
11. Rao, R.: Jaya: a simple and new optimization algorithm for solving constrained and unconstrained optimization problems. Int. J. Ind. Eng. Comput. **7**(1), 19–34 (2016)
12. Rao, R., Saroj, A.: Constrained economic optimization of shell-and-tube heat exchangers using elitist-jaya algorithm. Energy **128**(1), 785–800 (2017)
13. Rao, R., Saroj, A.: A self-adaptive multi-population based jaya algorithm for engineering optimization. Swarm Evol. Comput. **37**, 1–26 (2017)
14. Welch, G., Bishop, G.: An introduction to the kalman filter. University of North Carolina at Chapel Hill, Chapel Hill, NC, USA (1995)
15. Pérez, P., Gangnet, M., Blake, A.: Poisson image editing. ACM Trans. Graphics (TOG) **22**(3), 313–318 (2003)
16. Li, K., Yan, B., Wang, W., Gharavi, H.: An effective video synopsis approach with seam carving. IEEE Signal Process. Lett. **23**(1), 11–14 (2016)
17. Wang, Y., Jodoin, P.M., Porikli, F., Konrad, J., Benezeth, Y., Ishwar, P.: Cdnet 2014: An expanded change detection benchmark dataset. In: Proceedings of the IEEE Conference on Computer Vision and Pattern Recognition Workshops. pp. 387–394 (2014)
18. Fuentes, L., Velastin, S.: People tracking in surveillance applications. In: 2 ieee international workshop on performance evaluation of tracking and surveillance. PETS2001 (2001)

Author Index

A
Adhikary, Anwesha, 87

B
Bandyopadhyay, Soma, 109
Bandyopadhyay, Swagata, 179
Basak, Debadipta, 21
Bhattacharya, Saumik, 195
Bhowmick, Sutanni, 21
Biradar, Nagshettappa, 55

C
Chakraborty, Debasish, 179
Chakraborty, Papri, 87
Chakraborty, Riya, 11
Chanamallu, Srinivasa Rao, 135
Chanda, Pramit Brata, 145
Chatterji, B. N., 167, 185
Chatterjee, Soumyadip, 95
Choudhury, D., 219
Chowdhury, Anal Roy, 3

D
Das, Aheli, 101
Das, Dipnarayan, 77
Deyasi, Arpan, 3, 11, 87

G
Gadgay, Basavaraj, 55
Ghatak, Subhankar, 227
Ghosh, Ratul, 87
Ghosh, Sayantari, 195
Goswami, Bijoy, 21
Goswami, R., 219
Gowre, Sanjaykumar, 55
Gupta, Sumit, 77

H
Haldar, Arindam, 21
Hassan, Khondekar Lutful, 45

K
Kar, Avijit, 167
Kulkarni, Subhash, 155
Kumar, Ramesh, 31

M
Majumdar, Ivy, 167
Mallick, Subhasis, 65
Mandal, Jyotsna Kumar, 45, 109
Mandal, Surajit, 31
Mondal, Deepanwita, 11
Mudi, Prasenjit Kumar, 201

N
Naskar, Mrinal Kanti, 185

P
Patra, Anirban, 179
Paul, Aniruddha, 145
Paul, Aritra, 145
Paul, Goutika, 21

J. K. Mandal et al. (eds.), *Information, Photonics and Communication*,
Lecture Notes in Networks and Systems 79,
https://doi.org/10.1007/978-981-32-9453-0

R
Raj, Perugu Ananth, 123
Ramya, G., 155
Raychaudhuri, Anindita, 65
Roy, Krishnendu, 3
Rup, Suvendu, 227

S
Saha, Arijit, 101, 179
Saha, Manas, 185
Sarkar, Soumyadip, 101
Sarkar, Subir Kumar, 21
Singh, Shalini, 65
Sircar, Ankit, 65
Sonth, Mahesh V., 55

T
Thakur, S. S., 109
Thirumala, Srinivas, 135

CPSIA information can be obtained
at www.ICGtesting.com
Printed in the USA
LVHW081229271019
635468LV00006B/782/P

9 789813 294523